METAL CHALCO
BIOSENSOR

Woodhead Publishing Series in Electronic and Optical Materials

METAL CHALCOGENIDE BIOSENSORS

FUNDAMENTALS AND APPLICATIONS

Ali Salehabadi

Morteza Enhessari

Mardiana Idayu Ahmad

Norli Ismail

Banshi Dhar Gupta

Woodhead Publishing is an imprint of Elsevier
50 Hampshire Street, 5th Floor, Cambridge, MA 02139, United States
The Boulevard, Langford Lane, Kidlington, OX5 1GB, United Kingdom

ISBN: 978-0-323-85381-1 (print)
ISBN: 978-0-323-86003-1 (online)

For information on all Woodhead Publishing publications
visit our website at https://www.elsevier.com/books-and-journals

Publisher: Matthew Deans
Acquisitions Editor: Stephen Jones
Editorial Project Manager: Tom Mearns
Production Project Manager: Fizza Fathima
Cover Designer: Mark Rogers

Typeset by MPS Limited, Chennai, India
Transferred to Digital Printing 2023

Contents

CHAPTER

1

Introduction

1.1 History

As early as 1906, M. Cremer found and expressed that the concentration of an acid (liquid) is proportional to the electric potential. He deduced that this proportion occurred between parts of the fluid located on opposite sides of a glass membrane. This study was completed in 1909 when Søren Peder Lauritz Sørensen invented an electrode for pH measurements. Griffin and Nelson demonstrated an immobilization of the enzyme invertase on aluminum hydroxide and charcoal in 1922. After that, as early as 1956, Leland C. Clark fabricated the first biosensor for oxygen detection. This biosensor was later known as the *father of biosensors*, or *Clark electrodes*. Subsequently, in 1962, 1969, 1975, and 1992 several (basic) biosensors were discovered for the detection of bio-species. The field of biosensors is a broad area that covers several categories ranging from basic science to engineering and bioengineering. The term "sensor" has achieved great interest since just about 1999 and increased gradually up to date. This remarkable progress is reflected in scientific publications. Fig. 1.1 shows the number of published papers in ScienceDirect from 1991 to 2023 [1]. The number of published papers is gradually increased. Just after 2021 covering the principles of basic sciences with fundamentals of micro/nanotechnology, electronics, and applicatory medicine.

1.2 Basic knowledge and characteristics

(Bio) sensors possess certain static and dynamic attributes which require optimization to improve the performance of the (bio)sensor. In the coming chapters, these features will be discussed in more detail,

Metal Chalcogenide Biosensors
DOI: https://doi.org/10.1016/B978-0-323-85381-1.00004-0

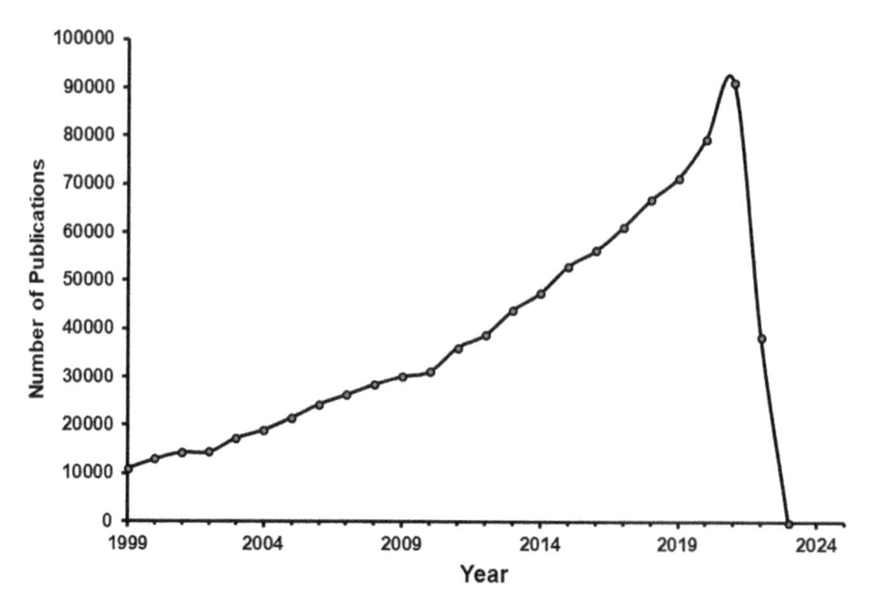

FIGURE 1.1 Number of publications versus year, reported in ScienceDirect.

such as selectivity, reproducibility, stability, sensitivity, and linearity [2]. All these features show one ability as;

Selectivity is the ability of a bioreceptor to detect a specific analyte in an admixture of various contaminants, such as selectivity between glucose and other carbohydrates, or antigen from the antibody. Selectivity is the main consideration when choosing receptors [3].

Reproducibility is the ability of a biosensor to create responses (100% identical) for a duplicated experimental set-up. The terms "precision" and "accuracy" are the major characteristics of any reproducibility. Reproducible signals show high reliability and robustness to the inference [4].

Stability is another ability of a biosensor which shows the degree of susceptibility to all disturbances around biosensing. The disturbances can cause errors in output signals. Stability is important for continuous monitoring such as thermal stability, chemical stability, etc [5].

Sensitivity is the ability of a biosensor which shows the "limit of detection" for measuring the minimum amount of analyte. For example, an environmental biosensor MUST be sensitive against ng/mL to fg/mL of environmental analytes [4].

And finally, linearity. It is the ability of a biosensor for accurate measurements of the measured response to a straight line as $Y = M \times C$, where Y, M, and C are the output signal, sensitivity of the biosensor, and concentration of the analyte, respectively. Linearity governs by the "resolution" and "range" of the analyte [6].

1.3 Overview of applications

Biosensors are devices for everyday life having a wide range of applications from the environment to health, foods, defense, and drugs. Detection of biomolecules is another crucial role of biosensors for indicating a disease in its early stage or controlling the known diseases. To date, various biosensors are fabricated for the detection of protein cancer biomarkers [7]. Food traceability, quality, safety, and nutritional value can also be controlled by biosensors. Environmental controlling of hazardous elements and gases can be another famous application of biosensors. Pollution monitoring, waste management, and industrial effluent control can be detected by appropriate biosensors. The biosensors require stability for long-term monitoring. The utilization of biosensors is known as one of the century's most important technological advancements for drug discovery, chemical, and biological detections, and toxic materials monitoring. In addition, biosensors are used (or under investigation) for use in prosthetic devices, and sewage epidemiology. The electrochemical, optical, and acoustic biosensors are utilized, along with their integration into analytical devices for various applications [6].

1.4 Nanotechnology in sensing devices

Nanotechnology is playing a progressively critical part in the advancement of biosensors. The affectability and execution of biosensors are being made strides by utilizing nanomaterials for their development. The utilization of these nanomaterials has permitted the presentation of numerous unused flag transduction advances in biosensors. Since their submicron measurements, nanosensors, nanoprobes, and other nanosystems have permitted basic and quick examinations in vivo. Versatile rebellious competent in analyzing numerous components are getting to be accessible. This work surveys the status of the different nanostructure-based biosensors. Utilization of the self-assembly procedures and nano-electromechanical frameworks (NEMS) in biosensors is talked about.

The utilization of nanomaterials in biosensors permits the use of numerous modern flag transduction advances in their fabrication. Since their submicron measure, nanosensors, nanoprobes, and other nanosystems are revolutionizing the areas of chemical and organic examination, to empower fast examination of different substances in vivo. Nanoparticles have various conceivable applications in biosensors. For illustration, utilitarian nanoparticles (electronic, optical and attractive) bound to natural particles (e.g., peptides, proteins, nucleic acids) have been created for utilization in biosensors to distinguish and intensify different signals. A few of the nanoparticle-based sensors incorporate acoustic wave biosensors, optical biosensors, and attractive and electrochemical biosensors, as examined following [8].

Acoustic wave biosensors, optical biosensors, magnetic biosensors, and electrochemical biosensors are the most important nano-based biosensors fabricated for various applications.

Acoustic wave biosensors are fabricated commonly using high-density nanoparticles such as Au, Pt, CdS, and TiO_2 and developed to greatly improve the sensitivity and limits of detection [9]. Optical biosensors are designed for recognizing specific DNA sequences. Gold nanoparticles have been utilized as widespread fluorescence quenchers to create an optical biosensor. This biosensor was created on this premise was able to identify single-base changes in a homogeneous organization [10]. Magnetic nanoparticles are an effective and flexible symptomatic apparatus in science and pharmaceutical. These materials ordinarily can be arranged within the shape of either a single space or superparamagnetic like Fe_3O_4, greigite (Fe_3S_4), and different sorts of ferrites ($MO \cdot Fe_2O_3$, where M = Ni, Co, Mg, Zn, Mn, etc.). Bound to biorecognition atoms, magnetic nanoparticles can be utilized to isolate or enhance the analyte to be recognized [11]. Electrochemical biosensors have been engineered from metallic nanoparticles. Metal nanoparticles can be utilized to upgrade the sum of immobilized biomolecules in the development of a sensor. Since of its ultrahigh surface zone, colloidal Au has been utilized to upgrade the DNA immobilization on a gold terminal, to eventually lower the local constraint of the manufactured electrochemical DNA biosensor [12].

In nano-biosensors, various classes of nanomaterials are used such as nano-wires, fibers, probes, tubular and porous nanostructures, chalcogenides, etc.

In summary, nanotechnology is an advancement in biosensors. Nanomaterials and nanofabrication advances are progressively being utilized to plan novel biosensors. The term "nano" is special to nanomaterials and is their most alluring perspective.

1.5 Challenges

Biosensors have been under improvement for around 50 years, a long time, and the inquiry into this field has produced a scholarly community over the final 10 years. In any case, other than sidelong stream pregnancy tests and electrochemical glucose biosensors, exceptionally few biosensors have accomplished worldwide commercial victory at the retail level. There are a few components for this: challenges in interpreting scholastic to investigate commercially practical models by industry; complex administrative issues in clinical applications; and it has not continuously been minoring to either discover analysts with a foundation in biosensor innovation or lock-in analysts from diverse disciplines of science to work together. Another reason is that scholastic investigation is driven by recommendations from the peer audit of science, financing organizations, and legislative issues that are in some cases characterized by different clashes of intrigue. It is regularly a jury of scholastics who decide the needs of financing offices with lawmakers who look for impressive warrants for the financing they endorse. In case a subject can be made to seem fancy and alluring, it encompasses a way better chance of victory. In this perspective, biosensor innovation includes a certain qualification that has been capably sold as a need. Biosensors ought to be pointed out as viable gadgets to be utilized. Although biosensors utilize crucial sciences, they can barely be thought of as "curiosity-driven" inquiries. On the other hand, inquiring about industry complies with s the slant of "follow the money" to a few degrees. Given the victory of commercial glucose sensors, biosensor inquiry is, of course, exceptionally profitable for the industry's long-term supportability. Be that as it may, it takes very a long time to deliver a commercially practical gadget from a verication of concept illustrated in the scholarly world [13]. This moreover includes managers that businesses are hesitant to confront. As a result, there are unaddressed required issues concerning the generation of a commercial biosensor, such as:

- Identification of the market
- Clear-cut advantages over existing methods for analyses of that analyte
- Testing the performance of the biosensor
- Response of a biosensor
- Stability, costs, and ease of manufacturing
- Hazards and ethics

The great news approximately biosensing advances is that most of the boundaries laid out are being broken quickly. High levels of speculation have been poured into translational inquiries around the world, especially, for healthcare applications. This brings the industry closer to the scholarly world to supply commercially practical products. On the other hand, there has been an exceptional advancement in the way researchers work over boundaries. Design and basic researchers these days have a much better understanding of fundamental biomolecular forms, whereas organic chemists and atomic scientists have more noteworthy mindfulness of the capabilities of diverse technologies. The alliance of specialists of diverse disciplines from the onset of biosensing advancement ventures may be an exceptionally appealing suggestion that will certainly bring progressed and novel items to the advertising.

1.6 Summary

Novel materials and technologies are urgently required for use in biosensors. Nanomaterials-based mixed metal oxides and metal chalcogenides in biosensors should be integrated within tiny biochips with onboard electronics. The quality of the biosensors must be moved forward to characterize the composition and rate constants related to atomic intelligence. Numerous artifacts related to authoritative information can be minimized or disposed of by planning the exploration legitimately, collecting information beneath ideal conditions, and preparing the information with reference surfaces. It is conceivable to universally fit high-quality biosensors with materials bimolecular response models, which approves the innovation as a biophysical apparatus for interaction analysis.

References

[1] N. Bhalla, P. Jolly, N. Formisano, P. Estrela, Introduction to biosensors, Essays Biochem. 60 (1) (2016) 1. Available from: https://doi.org/10.1042/EBC20150001.

[2] R. Karunakaran, M. Keskin, Biosensors: components, mechanisms, and applications, Anal. Tech. Biosci. ((2022) 179–190. Available from: https://doi.org/10.1016/B978-0-12-822654-4.00011-7.

[3] R. Guider, D. Gandolfi, T. Chalyan, L. Pasquardini, A. Samusenko, C. Pederzolli, et al., Sensitivity and Limit of Detection of biosensors based on ring resonators, Sens. Bio-Sens. Res. 6 (2015) 99–102. Available from: https://doi.org/10.1016/J.SBSR.2015.08.002.

[4] P.C. Pandey, S. Upadhyay, H.C. Pathak, C.M.D. Pandey, Sensitivity, selectivity, and reproducibility of some mediated electrochemical biosensors/sensors, Anal. Lett. 31 (14) (2006) 2327–2348. Available from: https://doi.org/10.1080/00032719808005310.

[5] N. Phares, R.J. White, K.W. Plaxco, Improving the stability and sense of electrochemical biosensors by employing trithiol-anchoring groups in a six-carbon self-assembled monolayer, Anal. Chem. 81 (3) (2009) 1095. Available from: https://doi.org/10.1021/AC8021983.

[6] A. Salehabadi, M. Enhessari, Application of (mixed) metal oxides-based nanocomposites for biosensors, Mater. Biomed. Eng. Inorg. Micro- Nanostruct. (2019). Available from: https://doi.org/10.1016/B978-0-08-102814-8.00013-5.

[7] A. Haleem, M. Javaid, R.P. Singh, R. Suman, S. Rab, Biosensors applications in the medical field: a brief review, Sens. Int. 2 (2021) 100100. Available from: https://doi.org/10.1016/J.SINTL.2021.100100.

[8] H. Kumar, K. Kuča, S.K. Bhatia, K. Saini, A. Kaushal, R. Verma, et al., Applications of nanotechnology in sensor-based detection of foodborne pathogens, Sensors (Basel, Switz.) 20 (7) (2020) 1996. Available from: https://doi.org/10.3390/S20071966.

[9] R. Fogel, J. Limson, A.A. Seshia, Acoustic biosensors, Essays Biochem. 60 (1) (2016) 101. Available from: https://doi.org/10.1042/EBC20150011.

[10] S.M. Borisov, O.S. Wolfbeis, Optical biosensors, Chem. Rev. 108 (2) (2008) 423–461. Available from: https://doi.org/10.1021/CR068105T.

[11] V. Nabaei, R. Chandrawati, H. Heidari, Magnetic biosensors: modelling and simulation, Biosens. Bioelectron. 103 (2018) 69–86. Available from: https://doi.org/10.1016/J.BIOS.2017.12.023.

[12] D. Grieshaber, R. MacKenzie, J. Vörös, E. Reimhult, Electrochemical biosensors - sensor principles and architectures, Sensors (Basel, Switz.) 8 (3) (2008) 1400. Available from: https://doi.org/10.3390/S80314000.

[13] M.L. Sin, K.E. Mach, P.K. Wong, J.C. Liao, Advances and challenges in the biosensor-based diagnosis of infectious diseases, Expert. Rev. Mol. Diagn. 14 (2) (2014) 225. Available from: https://doi.org/10.1586/14737159.2014.888313.

CHAPTER

2

Sensors and biosensors

2.1 Overview

Modern life is in direct relation to modern technologies. Modern technologies are changing every part of our lives, rapidly affecting our physical and mental health. These technologies have improved the everyday lives of many people. A simple and perspicuous example is a smartphone, have you imagined life without your smartphone? Though there are dual arguments about the technologies and life improvements, however, no one can ignore the role of technological advances in human health, particularly those in medical fields.

Sensing devices are one of the modern technological advancements in our lives with multidisciplinary applications ranging from industries to the environment and human health [1,2]. It is defined as any module or chip which can record the changes in the physical world and recover the readable responses. Sensors are traditionally classified under two categories: active and passive. "Active" sensors require an external excitation signal, while "passive" sensors can directly create an output response [3].

The sensors are also classified in terms of "detection used," "conversion mode," and "electrical output." The sensors-based detection used can be categorized under "materials detection" and "physical properties detection." Table 2.1 shows important examples of these two classes of sensors. In industries, the coalition of both homogeneous or heterogeneous sensors can produce more accurate results, particularly in terms of target detection, localization, and tracking as compared to a single sensor. Integrated sensors show a crucial role in chemical, biochemical, military, industry, environment, and health sense.

The next classification is based on conversion principles. In these sensors, one kind of energy converts to another kind of energy using

Metal Chalcogenide Biosensors
DOI: https://doi.org/10.1016/B978-0-323-85381-1.00010-6

TABLE 2.1 Various classifications of sensors, their detection, and conversion profiles.

Categories	Sub-classes	Sensor example	Detection target	Conversion profile
Based detection	Materials (gas, solid, liquid)	Gas	Atmosphere gases Toxic gases Explosive gases Volatile organic compounds Humidity Odor	Detecting the presence and concentration of various hazardous gases and vapors.
		Chemical	Organic Organometallic Inorganic	Transforming chemical information into a useful signal
		Environmental	Air Soil Water	Measuring air temperature, humidity, air quality, pressure, dust concentration, and noise
		Biological	Enzyme biosensor Tissue biosensor Immunosensor DNA biosensor	Self-containing analytical devices are comprised of a biological sensing element, or biorecognition element, which can convert the recognition phenomenon into a measurable signal.
		Radioactive	Alpha Beta Gama	Measuring nuclear, electromagnetic, or light radiation.
	Physical properties	Temperature sensor	Temperature	Measuring/monitoring the temperature to generate signal temperature changes.
		Proximity sensor	Object	Detecting the presence or absence of objects using electromagnetic fields, light, and sound without physical contact.
		Accelerometer	Acceleration (frequency, intensity, duration of physical activity)	Measuring the vibration, or acceleration of motion of a structure.

		IR sensor	IR	Measuring/detecting infrared radiation (IR) in its surrounding environment
		Pressure sensor	Gas pressure Touch pressure	Measuring the pressure
		Position sensor	Object Person Substance Electrical field	Converting physical parameters to an electrical output and measuring the
		Magnetic sensor (Hall effect sensor)	Magnetic field	Detecting the magnitude of magnetism and geomagnetism generated by a magnet or current.
		Microphone (sound sensor)	Sound	Measurement of how loud a sound is.
		Tilt sensor	Tilt orientation Tilt inclination	Measuring the tilt in multiple axes
		Flow and Level Sensor	Level sensing Point level detection of liquids	Determining the level or amount of fluids, liquids, or other substances that flow in an open or closed system
Based conversion		Photoelectrical	Photo-optical workpieces	Detecting objects, changes in surface conditions, and other items through a variety of optical properties
		Thermoelectrical	Heat transfers Low velocity Fluid flow	Measuring temperature differences to electric voltage
		Electrochemical	Composition of a system	
		Electromagnetic	Health monitoring Damage detection	Measurements of ambient (surrounding) electromagnetic fields
		Thermo-optical		
Electrical output		Analog Digital		

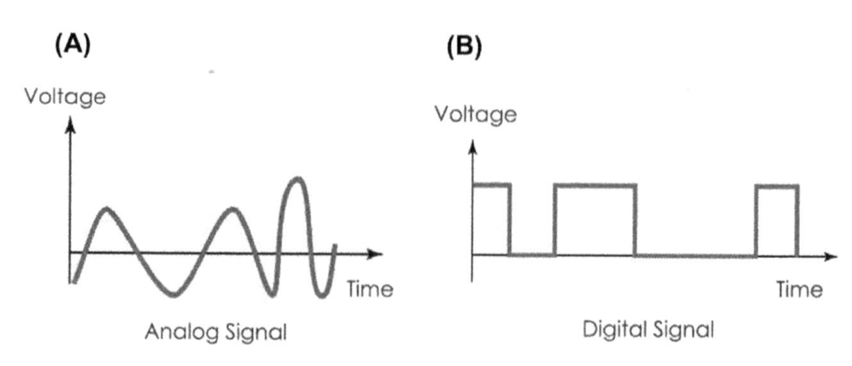

FIGURE 2.1 Comparison between analog and digital signals.

transducers generally via three steps: input device—process—output device [4]. For example, in a sound sensor (1) a microphone is an input device at which sound waves are converted into electrical signals for (2) a process of amplifying and (3) finally converting into an output device which can be a speaker [5]. Some of the common conversion phenomena are photoelectric, thermoelectric, electrochemical, electromagnetic, thermooptic, etc.

The most traditional classification of sensors is analog and digital sensors. An analog sensor produces analog output, that is, a continuous output signal, and generally senses the external parameters like wind speed, and solar radiation and gives analog voltage as an output, while a digital sensor works with discrete or digital data [6]. The data in digital sensors, which are used for conversion and transmission, is digital. Fig. 2.1 compares the schematic of a digital and an analog sensor.

As mentioned before, a sensor is a device that can convert an input quantity into an output signal. The output can be in the form of electrical or optical signals. Sensors and their associated interfaces are responsible for the detection of different physical and chemical properties of compounds such as temperature, pH, force, odor, pressure, the presence of special chemicals, flow, position, and light intensity [7,8].

Sensing devices are characterized as:

1. Merely sensitive to the chemical or physical quantity and no other parameters.
2. Does not influence the properties of the input chemical and/or physical quantities.

The affectability of a sensor demonstrates the degree of variation of the output relative to the alteration of the measured chemical or physical properties. The selection of a sensor should be based on such essential features as its selectivity, sensitivity, accuracy, calibration range,

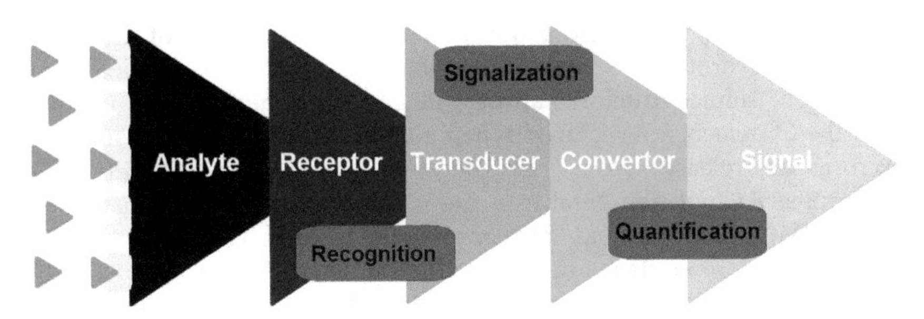

FIGURE 2.2 Sectional steps of a typical sensor.

resolution, cost-effectiveness, and repeatability as well as the prevailing environmental conditions.

The working profiles of a typical sensor include five sections from analyte (input) to signal (output) (Fig. 2.2). For example, five sections of a biosensor are [9]:

1. bio-analyte: a biological component of interest that needs detection, for example, glucose.
2. bioreceptor: a molecule that specifically recognizes the analyte such as an antibody, aptamer, enzyme, nucleic acid, etc.
3. bio-transducer: this part works based on conversion principles, and it is an element that converts one form of energy into another. This process is also called signalization.
4. converter: a part that processes the signalization of the transducer to signal for display.
5. detector or display: an interpretation system capable of detecting and identifying the processed data from the converter to the readable outputs such as numeric, graphic, tabular, or image.

A biosensor with all the above working sections offers superior advantages in terms of selectivity and sensitivity to the target analyte. The interaction between the bioreceptor and bio-analyte governs the workability of the sensors. In a biosensor, this interaction prevents the interference of signals from other substances with the desired biosensor signal.

Advanced materials, particularly solid-state inorganic materials, are developing rapidly for various applications ranging from small diodes to large photovoltaic panels. Though various applications of advanced materials have been reported, however, there has been a great increase in interest in the synthesis of novel multi-elemental inorganic materials such as mixed metal oxides (MMOs) [10], metal chalcogenides [11], alloys [12], metal-organic frameworks (MOFs) [13], and nanocomposites [14].

Several (mixed) metal oxides are semiconductors, where their properties are governed by a small change in stoichiometry and a small deficit of O atoms. Here, the electrons occupy the conduction band of the metal orbitals, instead of the localized O-atom orbitals [15]. The electrical conductivity of semiconductors can be changed upon heating and cooling since the O defects are replaced when atoms are added. As a result, the electrons are released from their sites (here conduction band) to form oxide ions. At high temperatures, the conductivity of semiconductors-based MMOs increases, as further oxygen is lost. This causes more defects and more electrons in the conduction band. This is d-type semiconduction and can be observed in high oxidation d-metal materials. The second group is p-type semiconduction materials including low oxidation number d-metal chalcogenides and halides. In p-type semiconduction, the process of electron generation occurs via a process equivalent to the oxidation of metal atoms. As a result, holes appear in the metal band, where the number of holes can increase upon heating. There is a direct relation between the number of holes and conductivity. The third types of semiconductor are n-type. In n-type semiconduction, the metal can be reduced to a lower oxidation state via the occupation of a conduction band formed from the metal orbitals [16,17].

Paramount advantages of solid-state materials for various applications including sensors are due to their structural and morphological flexibility, fabrication easiness, and costs of candidate materials. Various organic and inorganic materials till now have been profoundly investigated for sensing devices such as carbon materials, MMOs, MOFs, and chalcogenides. Among them, semiconductor metal chalcogenides (sulfide, selenide, and telluride) have received remarkable attention. A *chalcogenide* is a chemical compound comprised of at least one chalcogen anion and at least one electropositive element. Even though all group 16 elements of the periodic table are defined as chalcogens, the term chalcogenide is more frequently used for sulfides, selenides, and tellurides, rather than oxides. Metal chalcogenides show integrated chemical, optical, thermo-mechanical, and electrical properties. In addition, the mixed-metal chalcogenides are known for their optimal conversion efficiency, physicochemical stability, almost low band gap, and diverse crystal structures [18,19]. In the next coming sections, we will focus on various materials in the field of sensors, and important parameters in sensing measurements.

2.2 Sensing materials

Sensors respond to stimuli and use for either "identification" or "quantification" of various analytes such as chemicals, light, sound, and motion. In industries, sensors are strategic "controlling" components of manufacturing. The term "control" covers both "quality" and "online" processing. Physical sensors for controlling important industrial parameters (temperature, pressure, flow, etc.) have long been used [20]. However, in modern industries, chemical sensors are essential for the measurements of chemical species such as gas.

The working principles of physical and chemical sensors are different; a physical sensor is only influenced by a few parameters, while a chemical sensor is influenced by several parameters. In addition, most chemical sensors are expensive devices adjusted with complicated equipment and sampling methods, while most physical sensors involve simple materials and working methods [21].

Most chemical sensors detect analytes in gaseous or liquid phases. The chemical sensors are used in various fields ranging from detection of atmosphere pollution and hazardous laboratory gases to tracking explosive matters and hazardous chemicals in soil. In the health and medicine categories, the chemical sensors are adjusted and used for monitoring blood biochemical analyses and oxygen contents. A breath analyzer (breathalyzer) is a type of chemical sensor for detecting alcohol content, or diseases lung [22].

Most chemical sensors can be described using two important terms: criteria and characteristics. The major parameters of a chemical sensor are stability, repeatability, response time, and saturation. Moreover, sensitivity and selectivity are two crucial factors for detecting desired target species, particularly in a mixture of various chemical species. In the evaluation of chemical sensors, the qualification of selectivity is also important. It must be noted that a 100% selective sensor does not exist and there are always some interferences. The recent developments in the field of sensing materials have been collected in this section. These have been in turn categorized into (1) direct sensors, and (2) complex sensors [23].

Direct sensors utilize chemical reaction phenomena that directly affect a measurable electrical quantity such as resistance, potential, current, or capacitance. These devices do not require any energy conversion from one form of energy to another. On the other hand, the complex sensors employ chemically influenced phenomena that do not directly affect the electrical characteristics (e.g., physical shape change, frequency shifts, change in heat, etc.) and will require some transduction to obtain an electrical signal to interface with the common measurement electronics [24]. Some of the sensing materials encountered in direct sensors and complex sensors are given in Fig. 2.3.

2.3 Transduction elements

The fundamental transduction principles are used in sensors. Based on a general definition, the transducer is an electrical/electronic/electromechanical/electrochemical device that can convert energy 1 → energy 2

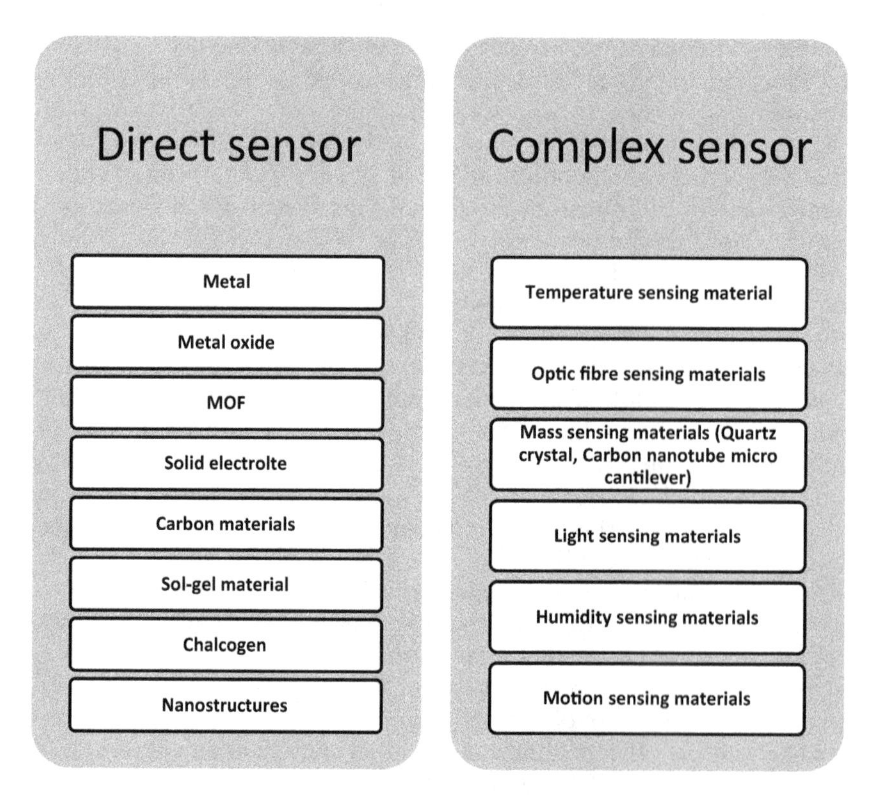

FIGURE 2.3 Sensing materials encountered in direct sensors and complex sensors.

based on the conversion principles. As a result of energy conversion, quantitative/qualitative measurements and also information transfer can occur. Transducers can be sensors or actuators with the ability to convert a signal into another [25].

Transducers are ordinarily planned to sense a particular measure and to, in a perfect world, react as it were to that specific measurand. A transducer will most likely react to the measurement and will moreover react to other energy sources that act on the sensor. For example, measuring strain with a piezoresistor where the main measurand is strain, while the resistance can also change with temperature. Common transduction techniques can be classified as:

- Piezoresistance—a mechanical input like pressure or force, which applies to a mechanical structure such as a beam or diaphragm. As a result, the structure to experience mechanical strain will obtain [26].
- Piezoelectricity—is a type of electroacoustic transducer that converts the electrical charges produced by some forms of solid materials into energy [27].

- Capacitive—a passive transducer that works and can measure physical quantities such as displacement, pressure, etc [28].
- Resistive—a resistive transducer is an electronic device that is capable of measuring various physical quantities like temperature, pressure, vibration, force, etc [29].
- Tunneling—tunneling displacement can detect the thermal expansion of a small volume of trapped gas. In quantum, the tunneling transducers are advanced nanoelectromechanical systems, which can be used for motion detection at the nanoscale [30].
- Thermoelectricity—thermoelectric transducer elements transform changes of the measured value such as temperature to change the current, arising from differences in temperature at the junction of two dissimilar materials. This is sometimes known as a thermocouple [31].
- Optical—a transducer that can convert light into electrical quantity, and also called photoelectric transducers [32].
- Radiation—a transducer that converts optical signal/light, into an electrical signal/current/voltage [33].
- Electrochemical—a transducer that shows changes in form of the electrical signal via binding of the analyte. This binding causes ionic discharge, which can be further recorded in the form of current/voltage [34].

2.4 Synthesis of solid-state materials for the sensing device

Various materials have been developed for the production of sensors. Synthesis and applications of nanoscale materials have been widely developed in everyday lives, owing to their improved structural, electrical, optical, magnetically, and chemical performances. Small sizes and large surface-to-volume ratios are two important features of nanostructuring. The major challenge of nanostructuring is exploring new methods with mild conditions and convenient operations. In this section, four important materials (mixed metal oxides (MMOs), metal chalcogenides, polymers, and nanocomposites) in the fabrication of sensing devices will be discussed in terms of "synthesis." In addition, we will answer this important question: why these materials are potential for sensor devices?

2.4.1 Mixed metal oxides

Two or more different kinds of metal cations in one composition with oxygen form MMOs. Based on the number of metal cations, the

MMOs are categorized as binary, ternary and quaternary, etc. In addition, the MMOs are sometimes classified as crystalline, and amorphous. In crystalline MMOs, the crystal structure is classified as:

- Perovskites with a chemical formula of ABO_3
- Scheelites, with a chemical formula of ABO_4
- Spinels with a chemical formula of AB_2O_4
- Palmeirites with a chemical formula of $A_3B_2O_8$.

In all the above chemical formulas, A and B are metal cations. The arrangement of cations differs by their coordination profiles and the nature of the neighboring cations. It is still not known, which of the constituent metal cations are active centers in mixed metal oxides. MMOs are widely used as industrial catalysts, which are usually multiphase systems. The activity of catalysts is governed by the presence of some phases.

There are two methods for the synthesis of MMOs, wet and dry chemical methods. Wet chemical synthesis is classified as a hydrothermal, solvothermal, sol-gel method, ultrasonic, coprecipitation, etc. are some renowned wet-chemical methods utilized for the preparation of highly crystalline MMOs. The dry chemical method (also called the solid-state chemical method) is the simplest method for the production of single crystal and polycrystalline MMOs, where high temperature ($500 < \text{temperature} < 2000°C$) is used for the formation of metal–oxygen–metal bonds. The dry method is kinetically and thermodynamically unfavorable [35]. The overall mechanism of a dry-chemical method for the synthesis of MMOs is summarized in Fig. 2.4.

2.4.2 Metal chalcogenides

A large family of 2D materials belongs to metal chalcogenides (MCs). MCs show several advanced features in energy conversion applications. These materials are classified into two main groups: (1) transition metal chalcogenides (TMCs), and (2) main group metal chalcogenides (MMCs).

TMCs are divided into two subclasses: transition metal dichalcogenides (TMDs with a general formula of MX_2) and transition metal tri-chalcogenides (TMTs with a general formula of MX_3).

$$MX_2 - \text{M:Mo, W and X:S, Se, Te}$$

$$MX_3 - \text{M:Ti, Zr, Hf and X:S, Se, Te}$$

Monolayer TMDs show an indirect to direct bandgap ranging from 1.0 to 2.1 eV, while monolayer TMTs cover the bandgaps in the range of 0.21 to 1.90 eV. Other combinations of M and X like MX, MX_2, and M_2X_3 are also promising candidates for optoelectronic applications [36].

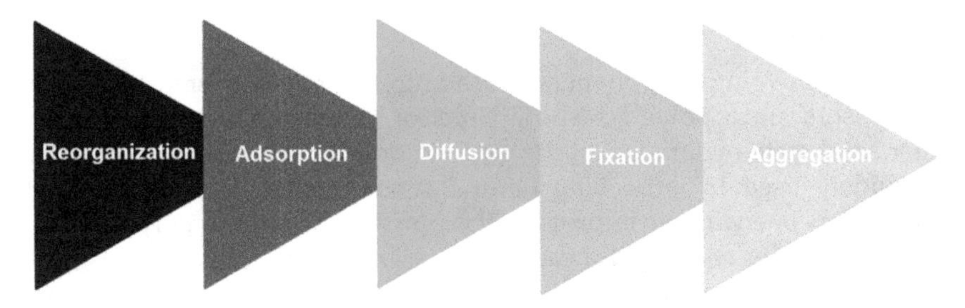

FIGURE 2.4 Five steps dry-chemical methods; reorganization - adsorption - diffusion - fixation (termination) - aggregation.

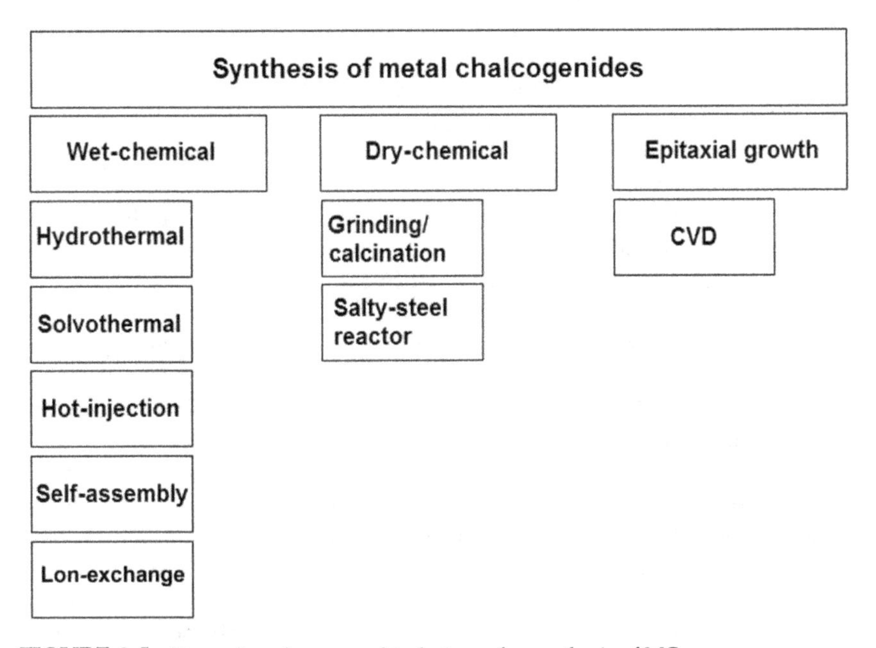

FIGURE 2.5 Examples of renowned techniques for synthesis of MCs.

Like nanoscales mixed metal oxides (MMOs), nanocrystalline metal chalcogenides can be synthesized via (1) solid-phase-, and (2) wet-chemical syntheses. Dry synthesis method is the simplest method in which after grinding of the precursors, the admixture is subjected to high temperature. The wet-chemical method is a method with several techniques such as microwave-assisted heating, sonochemical method, sono-electrochemical method, photochemical synthesis, irradiation method, microemulsion route, low-temperature solid-state synthesis, solvothermal technique, electrochemical template synthesis, and sol-gel method [37]. Fig. 2.5 shows some of the renowned techniques for the synthesis of MCs.

2.4.3 Polymers

The engineering of polymers is finding attraction in smart devices like sensors since polymers show superior properties as compared to other solid-state materials [38]. In the field of artificial sensors, polymers are widely used in mimicking natural sense organs. The blending of polymers is one way for improving the functionality of the polymers for better selectivity and rapid measurements. In addition, the composition of classical sensor materials with polymers including nanomaterials can also change the intrinsic or extrinsic functions of polymers. Various polymers including both conductive and nonconductive are utilized in sensors and biosensors. Polymers in sensor devices have two important roles: (1) take part in sensing or (2) freeze the component of analyte sensing.

As mentioned before, numerous polymers like polyaniline, polypyrrole, polytyramine, polyaminophenol, Nafion, polyisobutylene, poly [di (ethyleneglycol) adipate], poly[bis(cyanoallyl) polysiloxane], polydimethylsiloxane, polydiphenoxyphospha-zene, polychloroprene, poly [dimethylsiloxane-co-methyl-(3-hydroxypropyl)-siloxane]-g-poly (ethylene glycol) 3-aminopropyl ether, hydroxy-terminated polydimethylsiloxane, polystyrene beads, etc are used in the fabrication of sensors and biosensors.

Synthesis of polymer (polymerization) is the process of forming a long chain (or network) via covalently bounded small molecules (monomers). There are two methods for the synthesis of polymer, step- and chain-polymerizations. A step-growth polymerization is the bounding of monomers to form the first dimer, then trimer, oligomer, and finally macromolecule which is also called a polymer. Chain-growth polymerization refers to the process of polymer formation in which the growth of a polymer chain proceeds solely by a reaction between the monomer and active site on the polymer chain. A new active site forms at the end of each growth step. Radical polymerization is a famous example of chain polymerization which occurs in three main steps: initiation, propagation, and termination. Fig. 2.6 shows step and chain polymerizations.

Conductive polymers (CPs) are a renowned class of solid-state materials in sensing devices [39]. CPs are organic-based macromolecules with intrinsically high electrical conductivities with outstanding performances such as electrical, optical, high thermomechanical properties, and environmental stability. Conducting polymers can be synthesized via the following methods:

- Chemical oxidation
- Electrochemical polymerization
- Vapor phase synthesis
- Hydrothermal

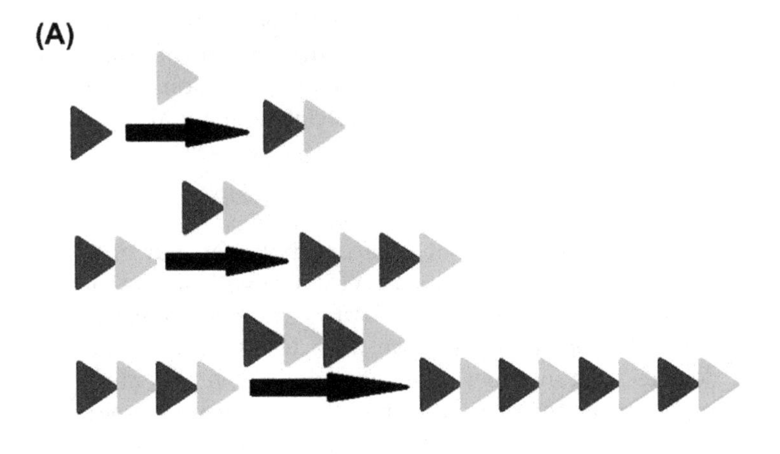

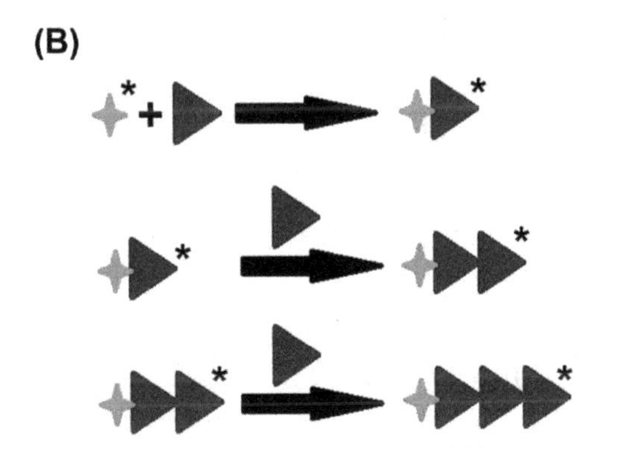

FIGURE 2.6 (A) step- and (B) chain- polymerizations; in step growth, the polymer chain doubles with each step, while in chain growth, the polymer chain always grows one monomer at a time.

- Solvothermal
- Electrospinning
- Self-assembly
- Photochemical methods
- Inclusion method
- Solid state method
- Plasma polymerization

Among them, chemical oxidation, and electro (co)polymerization methods are widely used for the synthesis of conductive polymers. In the chemical oxidation method, a monomer is mixed with an oxidizing agent in the presence of a suitable acid under ambient conditions to form a conductive polymer. Electrochemical polymerization is the process of deposition of the polymer onto the surface of solid electrode material through the mechanism of formation of cationic radical by the oxidation of the monomer on the solid electrode material.

2.4.4 Nanocomposites

Dula phase combination of at least one polymer and one inorganic material form polymer nanocomposites (PNCs) with several advantages in mechanical, electrical, and optical properties as compared to each compartment. Nanocomposites have attracted significant attention in sensing devices with minimal effort of various novel chemical and biological properties. Better selectivity, higher stability, and rapid measurements have been achieved by replacing classical sensor materials with polymers involving nanotechnology and exploiting either the intrinsic or extrinsic functions of polymers. Nanocomposites are classified under three categories:

1. Polymer matrix nanocomposites [40]
   a. Solution blending (solvent casting)
   b. Melt processing
   c. In-situ polymerization
2. Ceramic matrix nanocomposites [41]
   a. Mechanochemical
   b. Vapor phase reactions
   c. Elevated temperature synthesis
   d. Solution technique
3. Metal matrix nanocomposites [41]
   a. Liquid-phase
   b. Solid-phase
   c. Two-phase
   d. Deposition
   e. In-situ

2.5 Sensing measurement

Sensors provide an output signal in response to a physical, chemical, or biological measurement and typically require an electrical input to operate; therefore, they tend to be characterized in sensor specifications (commonly called "specs") or datasheets in much the same way as electronic devices. But the question is how sensors can measure analysts. To answer this important question, several factors and characteristics must be understood, depending on the sensing domain.

Sensor characteristics can be categorized under three main groups:

- *Systematic*: systematic characteristics as "those which can be exactly quantified by mathematical or graphical means.
- *Statistical*: The static accuracy of a sensor indicates how much the sensor signal correctly represents the measured quantity after it stabilizes (i.e., beyond the transient period.) Important static characteristics of sensors include sensitivity, resolution, linearity, zero drift and full-scale drift, range, repeatability, and reproducibility.
- *Dynamic*: The dynamic characteristics of a sensor represent the time response of the sensor system. Knowledge of these is essential to fruitfully using a sensor. Important common dynamic responses of sensors include rise time, delay time, peak time, settling time percentage error, and steady-state error.

Some of the most important characteristics of the sensors are presented in this section.

2.5.1 Range

The range is a static characteristic. Range implies the minimum and maximum values of the input or output. The term range is commonly used in the following ways in datasheets:

- A full-scale range (FSO) describes the maximum and minimum values of a measured property, often called "span." FSO is the algebraic difference between the output at maximum input and minimum input stimuli. Span values can be applied to a sensor with a high level of accuracy.
- The operating voltage range describes the minimum and maximum input voltages that can be used to operate a sensor. Applying an input voltage outside of this range may permanently damage the sensor.

2.5.2 Transfer function

Sensor characteristics illustrate the relevance between the physical quantity and the electrical output in the form of a signal. This relevance can demonstrate as a table, graph, or formula. The time-invariant of this relevance is known as the sensor transfer function. The transfer function can be expressed as; $S = F(x)$; where x and S are measurand and electrical signal, respectively. It is rare to find a transfer function that can be completely described by a single formula; therefore, functional approximations are used. The transfer function is classified into two sub-groups; linear and non-linear. Linearity is an expression of the extent to which the

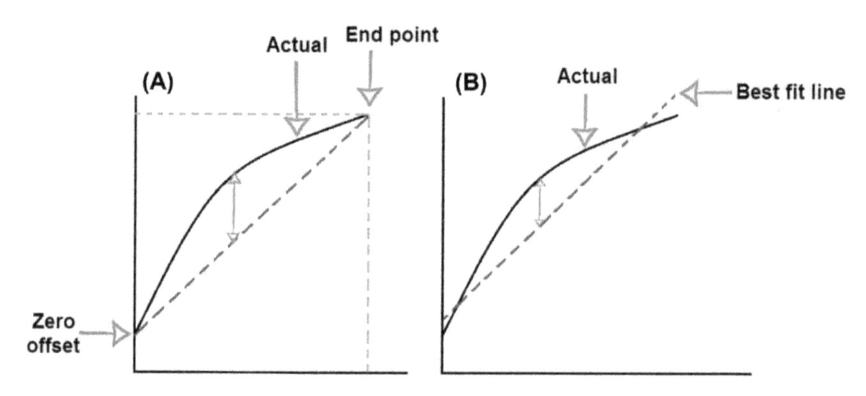

FIGURE 2.7 (A) end-point and (B) best-fit methods.

actual measured curve of a sensor departs from the ideal curve. Nonlinearity or linearity in datasheets is the difference between the actual line and the ideal straight line.

2.5.2.1 *Linear transfer functions*

A linear transfer function (LTFs) can be ascribed as; $S = A + Bx$; where A and B are the sensor's offset and slope, respectively. The sensor offset shows the output value of the sensor without measurand. The slope of an LTF illustrates "sensitivity." In practice, very few sensors are truly linear. However, it is considered to be linear characteristics when the plot of measurand versus the output values is approximately a straight line or linear approximation. The linear approximation can be measured graphically either by "end-point" or "best fit" methods (Fig. 2.7).

In the end-point method, which is the less accurate method, an ideal straight line is drawn between the upper- and lower-range values of the sensor. While in the best-fit method or independent linearity, the ideal straight line can be positioned in any manner that minimizes the deviations between it and the device's actual transfer function. The best fit method is the more accurate method and is commonly used by sensor manufacturers.

LTFs show several advantages such as:

- A simple method for calculating the measurand values from the electrical output
- The best method for predicting the electrical output is based on the measurand value.
- Easy to catch sensor offset and sensor slope characteristics.
- Accrue against non-ideal characteristics, such as nonlinearity and hysteresis.

Non-linear sensor output can be linearized utilizing software and hardware to use the preferences of sensors. The conventional hardware-based linearization strategy regularly requires manual calibration and exactness resistors to realize the required exactness. Cutting-edge smart sensors utilize less complex and less expensive advanced procedures to form a straight yield. These methods perform advanced linearization and calibration by leveraging the keen sensor's coordinates microcontroller and memory to store the plant calibration comes about for each person's sensor. The microcontroller can redress the sensor yield by looking for the recompense esteem or the real linearized yield in a look-up table. If memory is restricted, calibration coefficients, instead of a full look up the table, are utilized to build a linearized yield.

2.5.2.2 *Non-linear transfer functions*

When approximation does not well-matched to LTF, non-linear transfer functions (NLTFs) are used by approximation from other mathematical functions. Logarithmic functions; $S = A + B.\ln(x)$; exponential functions; $S = A.e^{k.x}$, and power functions; $S = A + B.x^k$. In these equations, x, S, A & B, and k are measurand, electrical signal, sensor parameters, like offset (for A) & slope (for B), and power factor, respectively. In addition, a polynomial function is another approximation and is used when none of the functions previously described can be applied. For sensor, polynomial approximations follow second- and third-order polynomials in the form of; $S = A \cdot x^2 + B \cdot x + C$; and $S = A \cdot x^3 + B \cdot x^2 + C \cdot x + D$, respectively. A third-order polynomial will provide a better fit as compared to a second-order polynomial, however, for a narrow range of input stimuli, it seems, that a second-order polynomial is more accurate.

As nonlinearity may vary along the input-output plot, a single value, called maximum nonlinearity, is used to describe this characteristic in datasheets. Maximum nonlinearity is typically expressed as a percentage of span. Nonlinearity can often be affected by environmental changes, such as temperature, vibration, acoustic noise level, and humidity. It is important to be aware of the environmental conditions under which nonlinearity is defined in the datasheet, particularly if they differ from the application operating environment.

2.5.3 Sensitivity

Sensitivity is the change in input required to generate a unit change in output. If the sensor response is linear, the sensitivity will be constant over the range of the sensor and is equal to the slope of the straight-line plot. An ideal sensor will have significant and constant sensitivity. If the sensor response is non-linear, the sensitivity will vary over the sensor range and can be found by calculating the derivative of S to x (dS/Dx).

2.5.4 Hysteresis

The output of a sensor may be different for a given input, depending on whether the input is increasing or decreasing. This phenomenon is known as *hysteresis* and can be described as the difference in output between the rising and falling output values for a given input. Like non-linearity, hysteresis varies along the input-output plot; thus, maximum hysteresis is used to describe the characteristic. This value is usually expressed as a percentage of the sensor span. Hysteresis commonly occurs when a sensing technique relies on the stressing of a particular material (as with strain gauges). Elastic and magnetic circuits may never return to their original start position after repeated use. This can lead to an unknown offset over time and can therefore affect the transfer function for that device.

2.5.5 Resolution

The resolution, also called discrimination, is the smallest increment of the measurand that causes a detectable change in output. The resolution of modern sensors varies considerably, so is important to understand the resolution required for an application before selecting a sensor. If the sensor resolution is too low for the application, subtle changes in the measurand may not be detected. However, a sensor whose resolution is too high for the application is needlessly expensive. A threshold is a name used to describe resolution if the increment is measured from zero, although it is more commonly described as the minimum measurand required to trigger a measurable change in output from zero.

2.5.6 Accuracy

Accuracy refers to a sensor's ability to provide an output close to the true value of the measurand. Specifically, it describes the maximum expected error between the actual and ideal output signals. Accuracy is often described relative to the sensor span. For example, a thermometer might be guaranteed to be accurate to within five percent of the span. As accuracy is relative to the true value of the measurand, it can be quantified as a percentage relative error using the following equation:

2.5.7 Precision

Precision is different from accuracy. Precision describes the ability of an output to be constantly reproduced. It is, therefore, possible to have a very accurate sensor that is imprecise (a thermometer that reports temperatures between 62°F and 64°F for an input of 63°F), or a very precise sensor that is inaccurate (a thermometer that always reports a temperature of 70°F for an input of 63°F). As precision relates to the reproducibility of a measure, it can be quantified as a percentage standard deviation using the following equation:

$$\text{Percentage standard deviation (PSD)} = \{(\text{standard deviation})/\text{mean}\} \times 100$$

2.5.8 Response time

Any changes in the input of sensors do not immediately reflect in the output. The period taken for the sensor to change its output from its previous state to a value within a tolerance band of the new correct value is called response time. The tolerance band is defined based on the sensor type, sensor application, or the preferences of the sensor designer. It can be defined as a value, such as 90% of the new correct value. Response time is commonly defined using time constants in first-order systems. A time constant is the time required by a sensor to reach 63.2% of a step change in output under a specified set of conditions. A time constant can be easily estimated by fitting a single exponential curve to the response curve.

2.6 Summary

Sensing devices are one of the modern technological advancements with several applications ranging from industries to the environment and human health. Any device which can record the changes in the physical world and convert them to readable responses is a sensor. Sensors are traditionally classified under two categories: Active and Passive. "Active" sensors require an external excitation signal, while "passive" sensors can directly create an output response.

References

[1] A. Salehabadi, N. Morad, M.I. Ahmad, A study on electrochemical hydrogen storage performance of β-copper phthalocyanine rectangular nanocuboids, Renew. Energy 146 (2020) 497–503. Available from: https://doi.org/10.1016/j.renene.2019.06.176.

[2] A. Salehabadi, M. Salavati-Niasari, M. Ghiyasiyan-Arani, Self-assembly of hydrogen storage materials based on multi-walled carbon nanotubes (MWCNTs) and Dy $3Fe_5O_{12}$ (DFO) nanoparticles, J. Alloy. Compd. 745 (2018) 789–797. Available from: https://doi.org/10.1016/j.jallcom.2018.02.242.

[3] M. Keerthika, D. Shanmugapriya, Wireless sensor networks: active and passive attacks - vulnerabilities and countermeasures, Glob. Transit. Proc. 2 (2) (2021) 362–367. Available from: https://doi.org/10.1016/J.GLTP.2021.08.045.

[4] M. Enhessari, A. Salehabadi, Perovskites-based nanomaterials for chemical sensors, Prog. Chem. Sens. (2016) 59–91. Available from: https://doi.org/10.5772/62559.

[5] K.A.A. Mamun, N. McFarlane, Integrated real time bowel sound detector for artificial pancreas systems, Sens. Bio-Sens. Res. 7 (2016) 84–89. Available from: https://doi.org/10.1016/J.SBSR.2016.01.004.

[6] S. Middelhoek, P.J. French, J.H. Huijsing, W.J. Lian, Sensors with digital or frequency output, Sens. Actuators 15 (2) (1988) 119–133. Available from: https://doi.org/10.1016/0250-6874(88)87002-1.

[7] W. Lu, Y. Guo, Y. Zhu, Y. Chen, Flexible sensors, Nanosens. Smart Manuf. (2021) 115–136. Available from: https://doi.org/10.1016/B978-0-12-823358-0.00006-X.

[8] P. Schagen, Some recent developments in remote sensing, Nature 266 (5599) (1977) 223–228. Available from: https://doi.org/10.1038/266223a0.

[9] A. Kawamura, T. Miyata, Biosensors, Biomater. Nanoarchitecton. (2016) 157–176. Available from: https://doi.org/10.1016/B978-0-323-37127-8.00010-8.

[10] K. Ramany, R. Shankararajan, K. Savarimuthu, S. Venkatachalapathi, G. Sivakumar, D. Murali, et al., Experimental verification of mixed metal oxide-based sensor for multiple sensing application, Mater. Lett. 301 (2021) 130248. Available from: https://doi.org/10.1016/J.MATLET.2021.130248.

[11] S. Kumar, V. Pavelyev, P. Mishra, N. Tripathi, P. Sharma, F. Calle, A review on 2D transition metal di-chalcogenides and metal oxide nanostructures based NO2 gas sensors, Mater. Sci. Semicond. Process. 107 (2020) 104865. Available from: https://doi.org/10.1016/J.MSSP.2019.104865.

[12] N. Kilinc, S. Sanduvac, M. Erkovan, Platinum-nickel alloy thin films for low concentration hydrogen sensor application, J. Alloy. Compd. 892 (2022) 162237. Available from: https://doi.org/10.1016/J.JALLCOM.2021.162237.

[13] J. Zhang, L. Gao, Y. Zhang, R. Guo, T. Hu, A heterometallic sensor based on Ce@Zn-MOF for electrochemical recognition of uric acid, Microporous Mesoporous Mater. 322 (2021) 111126. Available from: https://doi.org/10.1016/J.MICROMESO.2021.111126.

[14] H. Lin, F. Wang, Y. Duan, W. Kang, Q. Chen, Z. Xue, Early detection of wheat Aspergillus infection based on nanocomposite colorimetric sensor and multivariable models, Sens. Actuators B Chem. 351 (2022) 130910. Available from: https://doi.org/10.1016/J.SNB.2021.130910.

[15] J.L. Zhang, G.Y. Hong, Nonstoichiometric compounds, Mod. Inorg. Synth. Chem. Second. (Ed.) (2017) 329–354. Available from: https://doi.org/10.1016/B978-0-444-63591-4.00013-6.

[16] B. El Filali, O.Y. Titov, Y.G. Gurevich, Physics of charge transport in metal–monopolar (n- or p-type) semiconductor–metal structures, J. Phys. Chem. Solids 118 (2018) 14–20. Available from: https://doi.org/10.1016/J.JPCS.2018.02.047.

[17] D. Qu, T. Qi, H. Huang, Acceptor–acceptor-type conjugated polymer semiconductors, J. Energy Chem. 59 (2021) 364–387. Available from: https://doi.org/10.1016/J.JECHEM.2020.11.019.

[18] Q. Si, R. Yu, E. Abrahams, High-temperature superconductivity in iron pnictides and chalcogenides, Nat. Rev. Mater. 1 (4) (2016) 1–15. Available from: https://doi.org/10.1038/natrevmats.2016.17.

[19] B.J. Eggleton, B. Luther-Davies, K. Richardson, Chalcogenide photonics, Nat. Photon. 5 (3) (2011) 141–148. Available from: https://doi.org/10.1038/nphoton.2011.309.

[20] T.Y. Kim, W. Suh, U. Jeong, Approaches to deformable physical sensors: Electronic vs iontronic, Mater. Sci. Eng. R: Rep. 146 (2021) 100640. Available from: https://doi.org/10.1016/J.MSER.2021.100640.

[21] Jří. Janata, Principles of chemical sensors, Princ. Chem. Sens. (2009). Available from: https://doi.org/10.1007/B136378.

[22] L. Lindberg, D. Grubb, D. Dencker, M. Finnhult, S.G. Olsson, Detection of mouth alcohol during breath alcohol analysis, Forensic Sci. Int. 249 (2015) 66–72. Available from: https://doi.org/10.1016/J.FORSCIINT.2015.01.017.
[23] M.J. McGrath, C.N. Scanaill, Sensing and sensor fundamentals, Sens. Technol. (2013) 15–50. Available from: https://doi.org/10.1007/978-1-4302-6014-1_2.
[24] M.M. Arafat, A.S.M.A. Haseeb, A. Akbari, Developments in semiconducting oxide-based gas-sensing materials, Compr. Mater. Process. (2014) 205–219. Available from: https://doi.org/10.1016/B978-0-08-096532-1.01307-8.
[25] B.R. Eggins, Transduction elements, Anal. Tech. Sci. (2007) 11–67. Available from: https://doi.org/10.1002/9780470511305.CH2.
[26] S. Kouchakzadeh, K. Narooei, Simulation of piezoresistance and deformation behavior of a flexible 3D printed sensor considering the nonlinear mechanical behavior of materials, Sens. Actuators A: Phys. 332 (2021) 113214. Available from: https://doi.org/10.1016/J.SNA.2021.113214.
[27] T. Košir, J. Slavič, Single-process fused filament fabrication 3D-printed high-sensitivity dynamic piezoelectric sensor, Addit. Manuf. (2021) 102482. Available from: https://doi.org/10.1016/J.ADDMA.2021.102482.
[28] F. Abdollahi-Mamoudan, S. Savard, C. Ibarra-Castanedo, T. Filleter, X. Maldague, Influence of different design parameters on a coplanar capacitive sensor performance, NDT E Int. 126 (2022) 102588. Available from: https://doi.org/10.1016/J.NDTEINT.2021.102588.
[29] T. Dong, Y. Gu, T. Liu, M. Pecht, Resistive and capacitive strain sensors based on customized compliant electrode: comparison and their wearable applications, Sens. Actuators A: Phys. 326 (2021) 112720. Available from: https://doi.org/10.1016/J.SNA.2021.112720.
[30] J. Yu, C. Yue, C. Jiang, D. Zhang, X. Huang, C. Yang, et al., Research on suppression of external magnetic field interference of tunnel magnetoresistive sensor based on versoria variable step improved adaptive filtering method, Energy Rep. 7 (2021) 300–311. Available from: https://doi.org/10.1016/J.EGYR.2021.08.043.
[31] K. Wan, Z. Liu, B.C. Schroeder, G. Chen, G. Santagiuliana, D.G. Papageorgiou, et al., Highly stretchable and sensitive self-powered sensors based on the N-Type thermoelectric effect of polyurethane/Nax(Ni-ett)n/graphene oxide composites, Compos. Commun. 28 (2021) 100952. Available from: https://doi.org/10.1016/J.COCO.2021.100952.
[32] W. Zheng, Y. Zhang, L. Li, X. Li, Y. Zhao, A plug-and-play optical fiber SPR sensor for simultaneous measurement of glucose and cholesterol concentrations, Biosens. Bioelectron. (2022) 113798. Available from: https://doi.org/10.1016/J.BIOS.2021.113798.
[33] G.V. Selicani, F. Buiochi, Stepped-plate ultrasonic transducer used as a source of harmonic radiation force optimized by genetic algorithm, Ultrasonics 116 (2021) 106505. Available from: https://doi.org/10.1016/J.ULTRAS.2021.106505.
[34] A. Shaver, N. Arroyo-Currás, The challenge of long-term stability for nucleic acid-based electrochemical sensors, Curr. Opin. Electrochem. (2022) 100902. Available from: https://doi.org/10.1016/J.COELEC.2021.100902.
[35] M.A. Carreon, V.V. Guliants, Novel macroporous vanadium-phosphorus-oxides with three-dimensional arrays of spherical voids, Stud. Surf. Sci. Catal. 141 (2002) 309–316. Available from: https://doi.org/10.1016/S0167-2991(02)80557-2.
[36] M. Ates, E. Yılmaz, M.K. Tanaydın, Challenges, novel applications, and future prospects of chalcogenides and chalcogenide-based nanomaterials for photocatalysis, Chalcogenide-Based Nanomater. Photocatal. (2021) 307–337. Available from: https://doi.org/10.1016/B978-0-12-820498-6.00014-7.
[37] M.D. Khan, M. Opallo, N. Revaprasadu, Colloidal synthesis of metal chalcogenide nanomaterials from metal–organic precursors and capping ligand effect on electrocatalytic

performance: progress, challenges and future perspectives, Dalton Trans. 50 (33) (2021) 11347–11359. Available from: https://doi.org/10.1039/D1DT01742J.

[38] S. Cichosz, A. Masek, M. Zaborski, Polymer-based sensors: a review, Polym. Test. 67 (2018) 342–348. Available from: https://doi.org/10.1016/J.POLYMERTESTING.2018.03.024.

[39] Jiri Janata, M. Josowicz, Conducting polymers in electronic chemical sensors, Nat. Mater. 2 (1) (2003) 19–24. Available from: https://doi.org/10.1038/nmat768.

[40] R. Das, A.J. Pattanayak, S.K. Swain, Polymer nanocomposites for sensor devices, In: Polymer-Based Nanocomposites for Energy and Environmental Applications, Woodhead Publishing (2018) 205–218. Available from: https://doi.org/10.1016/B978-0-08-102262-7.00007-6.

[41] I. Graz, M. Krause, S. Bauer-Gogonea, S. Bauer, S.P. Lacour, B. Ploss, et al., Flexible active-matrix cells with selectively poled bifunctional polymer-ceramic nanocomposite for pressure and temperature sensing skin, J. Appl. Phys. 106 (3) (2009) 034503. Available from: https://doi.org/10.1063/1.3191677.

CHAPTER

3

Chalcogenides for sensing

3.1 Introduction

Various classes of inorganic materials are invented and synthesized with desired properties for energy applications either storage or conversion [1]. Semiconductors based metal oxides and metal chalcogenides particularly sulfide, selenide, and telluride have received remarkable attention in modern technologies, owing to their physical and chemical properties. In addition, these materials individually or in combination with other materials can impart remarkable thermal, electrical, and mechanical properties [2]. In the same vein, the chalcogenides are known as an optimal combination of decent conversion efficiency, ability to grow and deposit in ambient conditions, low band gap, band gap engineering, diverse crystal structures, nature to grow in layer forms, etc. In their composition, there are at least one chalcogen anion (A^{x-}) and at least one more electropositive element (B^{y+}). Even though all group 16 elements of the periodic table are defined as chalcogens, the term chalcogenide is more frequently used for sulfides, selenides, and tellurides, rather than oxides.

Transition metal chalcogenides nanostructures demonstrate various physical and chemical properties due to their size and shapes. Physically, it is known that the atoms at the surface have surrounded a smaller number of neighbors, therefore have higher average binding energy per atom. Moreover, atoms at the edges and corners have lower coordination potential, where these atoms can strongly bind with incoming atoms or molecules. Metal chalcogenides, particularly in nanoscales, have several intrinsic properties, which make them capable in energy-harvesting devices. The properties of chalcogenide nanostructures are sturdily reliant on method of synthesis, shape, size, crystallinity, nature of surface, presence of defects, etc. Though, there are several

Metal Chalcogenide Biosensors
DOI: https://doi.org/10.1016/B978-0-323-85381-1.00006-4

essential features of metal chalcogenide nanostructures, however, there are few research articles in the field of imperative metal chalcogenide nanostructures in terms of "synthesis," and "novel application."

In the following sections, we will discuss the detailed chemistry of the chalcogens and metal chalcogenides with respect to their unique performances in physical, optical, thermo-optical properties.

3.1.1 The chalcogens; an overview

A mentioned before, chalcogenides combine elements in group 16 of the periodic table (such as sulfur, selenium, tellurium, polonium, and livermorium) with more electropositive elements such as silver.

Sulfur (Z = 16) is a chemical element that is represented with the chemical symbol "S" and the atomic number 16 on the periodic table. Because it is 0.0384% of the Earth's crust, sulfur is the 17th most abundant element following strontium. Sulfur also takes on many forms, which include elemental sulfur, organo-sulfur compounds in oil and coal, $H_2S(g)$ in natural gas, and mineral sulfides and sulfates.

Selenium (Z = 34) was discovered by Swedish chemist Jons Jacob Berzelius in 1817. Selenium (Se) is a non-metal and can be compared chemically to its other non-metal counterparts found in Group 16: The Oxygen Family, such as sulfur and tellurium.

Tellurium (Z = 52) was discovered by von Reichenstein in 1782. It is a brittle metalloid that is relatively rare. It is named after the Latin tellus for "earth." Tellurium (Te) can be alloyed with some metals to increase their machinability and is a basic ingredient in the manufacture of blasting caps. Elemental tellurium is occasionally found in nature but is more often recovered from various gold ores.

Polonium (Z = 84) was discovered in 1898 by Marie Curie with the chemical symbol of "Po." The discovery was made by extraction of the remaining radioactive components of pitchblende following the removal of uranium. Current production for research purposes involves the synthesis of the element in the lab rather than its recovery from minerals. This is accomplished by producing Bi-210 from the abundant Bi-209.

Livermorium (Z = 116) was approved by IUPAC in 2012. The new name honors Lawrence Livermore National Laboratory (1952). A group of researchers of this Laboratory with the heavy element research group of the Flerov Laboratory of Nuclear Reactions took part in the work carried out in Dubna on the synthesis of superheavy elements including element 116.

The soft chalcogens S, Se, and Te form binary compounds with metals that commonly have quite different structures from the corresponding oxides, nitrides, and fluorides. This difference is consistent with the greater covalence of the compounds of sulfur and its heavier congeners. For example, it is known that the metal-oxygen (MO) compounds adopt the rock-salt structure, whereas ZnS and CdS can crystallize with either of the sphalerite or the wurtzite structures in which the lower coordination numbers indicate the presence of directional bonding. Similarly, the d-block monosulfides generally adopt the more characteristically covalent nickel-arsenide structure rather than the rock-salt structure of alkaline-earth oxides such as MgO. Even more striking are the layered Ms_2 compounds formed by many d-block elements in contrast to the fluorite or rutile structures of many d-block dioxides. Before we discuss these compounds, we should note that many metal-rich compounds that disobey simple valence rules and do not conform to an ionic model. Some examples are Ti_2S, Pd_4S, V_2O, Fe_3N, and even the alkali metal suboxides, such as Cs_3O. The occurrence of metal-rich phases is associated with metal–metal (M–M) interactions. Many other intermetallic compounds display stoichiometries that cannot be understood in terms of conventional valence rules. Included among them are the important permanent-magnet materials $Nd_2Fe_{17}B$ and $SmCo_5$.

Several characteristics of chalcogenides are similar to other mixed (complex) materials, then what does make chalcogenides different? This is an important question, which will be discussed in the next section.

3.1.2 What does make chalcogenide different?

Chalcogenide materials behave differently from oxides, owing to their lower band gaps and reactivity contributing to very dissimilar structural, thermal, optical, and electrical properties. Considering the significant covalence, the electrical conductivities, and the chemical properties, the electronic structure of the chalcogenides needs to invoke more on their band model (Fig. 3.1). In this figure, the band structures

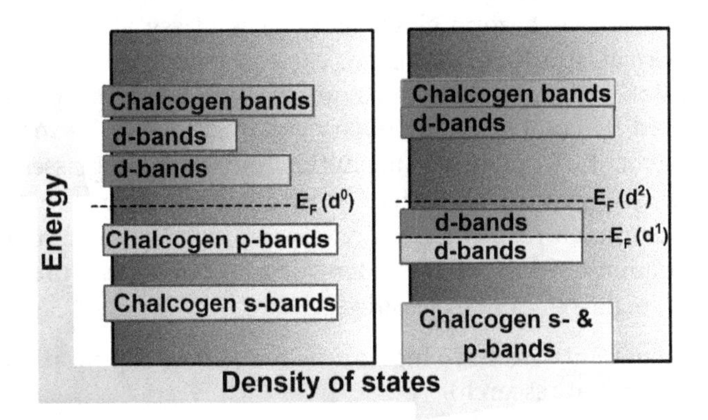

FIGURE 3.1 Approximate band structures of dichalcogenides.

are presented where the dichalcogenides with octahedral and trigonal prismatic metal sites have low-lying bands composed primarily of chalcogen s- and p-orbitals, with higher-energy bands from metal d-orbitals.

For example, soft chalcogens have rich halogen chemistry. Sulfur forms very unstable iodides, but the iodides of Te are more robust, which is an example of a large anion stabilizing a large cation. Among halogens, fluorine (F) shows the maximum group oxidation state of the chalcogen elements, but the low oxidation state fluorides of Se and Te are unstable with respect to disproportionation into the element and higher oxidation. The above pure chemistry explains just one part of the superior differences of chalcogens as compared to other materials.

Chalcogenide glass is a class of chalcogens that covalently bonded to form "covalent network solids" [3]. Chalcogenides glassy materials have wide optical windows in the mid-infrared region. This allows us to discover the vibrational modes of biomolecules. The optimal rheological properties of these glasses allow fiber drawing without devitrification and optical loss. They can be shaped and processed to reduce the diameter of the sensing zone for higher detection sensitivity and better mechanical flexibility. In addition, amorphous chalcogenides effectively resist chemical corrosion, which is good for biocompatibility with live biological components. The glassy chalcogenides have also an apparent hydrophobic surface characteristic, which appears to attract non-polar organic species while repelling water. Consequently, the optical signal of organic species is enhanced relative to water [4].

Chalcogenide glass-based devices have also shown the possibility of fabrication of ultralow loss waveguides due to their larger refractive index amongst glasses. Sulfur, selenium, and tellurium have generated significant interest over the past decade owing to their transparent behavior in the near-to mid-infrared region. Silver (Ag), copper (Cu), Iron (Fe), Cadmium (Cd), and Zinc (Zn) are some elements that can be used in combination with soft chalcogens to form glassy alloys. The glassy alloys show higher order of thermal, structural, morphological, and mechanical stability.

In addition to binary chalcogens, ternary and quaternary glasses are also reported by combining the binary compositions or by the addition of other elements. Comparing the sulfide and selenide glasses, the sulfide glasses are typically characterized by a wider bandgap, higher glass transition temperature, lower density, greater hardness, and higher thermal conductivity. Some of the interesting technological properties of chalcogenide glasses can be classified as:

1. Chalcogenide glasses have high refractive indices (application in photonic crystal research).
2. Chalcogenides glassed have relatively high atomic masses and weak bond strengths. These two properties cause low characteristic

phonon energies in the range of 350–450 cm^{-1}, narrow bandgaps ~2 eV, and relatively poor transparency in the visible and UV wavelength regions.

3. Chalcogenide glasses exhibit an array of photoinduced structural changes. It means light can induce almost a large change in glass absorption. Photodarkening and photobleaching are two phenomena that cover the term "glass absorption."
4. Chalcogenides can dissolve metals due to the unique photodoping of the chalcogens.
5. Chalcogenides have a reversible amorphous to crystalline phase change. This can induce by controlled thermal cycling (application in rewritable CD/DVD).
6. Chalcogenide glasses exhibit high nonlinearities in the near IR wavelength region, with two to three orders of magnitude higher than those of other glasses like silicate (application in optical devices in fiber networks).
7. Chalcogenide glasses rely on a nonlinear conduction mechanism. Here, the conductive filaments formed within the storage material by applying field (application in novel non-volatile memory technologies).

In addition to the physical and chemical properties of chalcogens, the soft chalcogens play a crucial role in biology such as cellular redox homeostasis and cysteine-derived antioxidants, whereas the heavier chalcogens like Selenium and Tellurium illustrate more antioxidant performance. These materials are capable of quenching peroxyl radicals in a catalytic fashion and express unusually high reactivity.

Since the unique properties of chalcogens are due to the chemistry and compositions of these compounds, it is crucial to study the structural properties of these materials. In the next section, we will focus on the structures and functional units of chalcogens.

3.1.3 Structures and functional units

As mentioned before, chalcogens are generally alloyed with electropositive elements. In this alloying system, Arsenic (As) and Germanium (Ge), and with elements like Phosphor (P), Antimony (Sb), Bismuth (Bi), Silicon (Si), Tin (Sn), Lead (Pb), Aluminum (Al), Gallium (Ga), Silver (Ag), Lanthanum (La), etc. to form chalcogenide glasses that exhibit fascinating properties.

The synthesis of chalcogenides is normally based on the inherent preference for a certain growth direction or the surfactant-driven shape guiding. Soft chalcogens (i.e., S, Se, Te) are representative material species having the preferred directional growth. The shape of nanocrystals is known to be dominated by the total minimum surface energy of the crystal facets; this is known as the Gibbs-Wulff theorem Venables.

In many chalcogenides, the minimized total surface energies are found in the 1D and 2D shapes because of the asymmetric bond strength. Se and Te are well known to grow into nanowires or nanotubes. The layer-structured chalcogenides grow into 2D nanoplates or nanosheets, as frequently observed in Bi_2Se_3, Bi_2Te_3, MoS_2, SnS_2, etc. The assembly technique has been developed in parallel with the synthetic advance to achieve an ordered arrangement in 2D structures or heterostructures with the aim of the realization of devices.

One-dimensional (1D) includes nanowire [5], nanorod [6], nanotube [7], nanobelt [8], nanoneedle [9], nanoribbon [10], nanofiber [11], and whisker [12]. For the formation of an anisotropic nanostructure, the crystal grows along a certain orientation faster than the others. Especially, nanowires with uniform diameter can be obtained when crystal grows along one direction with no or suppressed growth along the other directions. Periodic bond chain (PBC) theory explains the thermodynamical equilibrium crystal based on different growth rates (surfaces with high surface energy grow faster and disappear finally). Thus, only surfaces with the lowest total surface energy will survive. Thus, the formation of 1D nanostructures depending on the different growth rates is limited to some materials with special crystal structures. Furthermore, a low supersaturation is essential for anisotropic growth, otherwise, secondary or homogeneous nucleation can occur at high supersaturation. Nevertheless, numerous 1D metal chalcogenides have been synthesized by modulating the surface energy with surfactant or attachment. In addition, intrinsically grown 1D chalcogens or metal chalcogenides can exploit chemical or physical templates to produce various metal chalcogenides through chemical transformation. The reported synthetic protocols can be classified into four different pathways: (1) Intrinsic growth, (2) Shape-guiding agent growth, (3) Oriented attachment, and (4) Chemical transformation.

A family of 2D materials includes single elements such as silicene, germanene, phosphorene, boron nitride, metal oxides, and metal chalcogenides. High mechanical flexibility and transparency of ultrathin 2D nanomaterials are believed to bridge the microscopic properties to the emerging macroscopic features with the quantum confinement from the reduced dimensionality. In addition to mechanical exfoliation and CVD, wet chemistry proved its availability to obtain 2D metal chalcogen nanocrystals with high quality through liquid exfoliation, ion intercalation/exfoliation, solution-phase deposition, and chemical transformation.

Metal chalcogenides are categorized under three subclasses, based on their structures: binary, ternary, and quaternary. Other classification cations (again based on structures), transition metal chalcogenides are

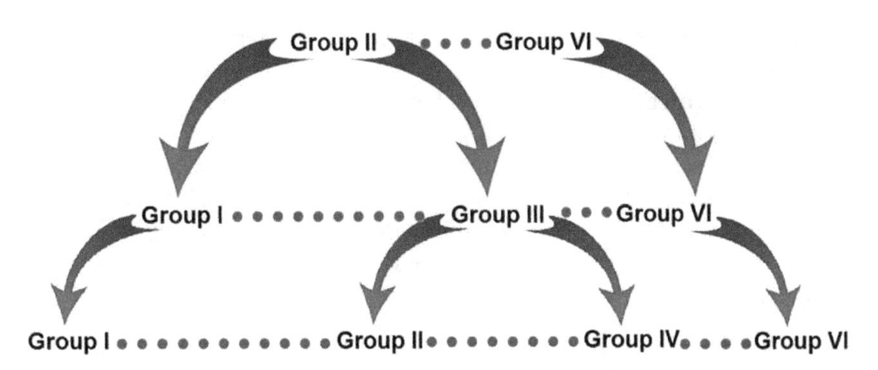

FIGURE 3.2 Relation between semiconductors for production of MCs; group II–VI, I–III–VI and I–II–IV–VI elements.

classified as monochalcogenides (MC, where M is a transition metal and C, is S, Se, Te), dichalcogenides (MC_2, where M is a transition metal and C, is S, Se, Te), tri- and tetrachalcogenides ($M^{4+}C_2^{2-}C^{2-}$ where M is generally Ti, V, Cr, Mn, and C is S, Se, Te).

Metal chalcogenides form at least one metal ion from Compounds comprising two groups VI elements (e.g., CuInSSe) or two group III elements (e.g., CIGS) will here be referred to as "pseudo-quaternary" systems. They derive from ternary semiconductors in a way that cations or anions are substituted with ions of the same oxidation state. This is not the case for semiconductors like Cu_2ZnSnS_4 (CZTS) and Cu2ZnSnSe4 (CZTSe), which we classify as quaternary compounds. Like ternary I–III–VI2 semiconductors can be conceptually derived from binary II–VI semiconductors by replacing two divalent cations by one monovalent and one trivalent cation, quaternary I2–II–IV–VI4 compounds can be derived from ternary I–III–VI2 semiconductors by substituting two trivalent cations by one divalent and one tetravalent cation (Fig. 3.1). CZTS and CZTSe are emerging materials, which have similar band gaps, p-type conductivity and absorption coefficients as $CuInS_2$ and $CuInSe_2$. Their main advantage is that they do not contain indium whose natural resources are prognosticated to become scarce in the soon. In the latter classification (Fig. 3.2), the elements from groups II and VI can form binary MCs, for example, InSe. Ternary MCs can form from the bonding of groups I, III, and VI, which are derivatives of binary II–VI. In this case, bivalent cations are substituted by mono and trivalent cations. In the same vein, quaternary MCs can be generated from group I–II–IV–VI elements, which are derivatives of ternary MCs of I–III–VI. This occurs by the replacement of trivalent cations (M^{3+}) with bivalent (M^{2+}), and tetravalent cations (M^{4+}). Let's see one example of this conversion: derivation of $Cu^{(I)}{}_2Zn^{(II)}Sn^{(IV)}S^{(VI)}{}_4$, from $Cu^{(I)}In^{(III)}S^{(VI)}{}_2$ and $Zn^{(II)}S^{(VI)}$. Here ZnS and $CuInS_2$ are primary and secondary parents, respectively. Fig. 3.2 clearly shows these relationships.

To date, numerous efforts have been developed to design complex metal chalcogenides for novel applications such as energy and sensor. In this field, several types of binaries, ternary, and quaternary metal chalcogenides are engineered and utilized for several applications. In this section, for the first time, we classify the binary, ternary, and quaternary metal chalcogenides based on their composition (or elemental compositions) as:

1. Binary metal chalcogenides
   a. Transition Metals (Table 3.1)
   b. Post Transition Metals (Table 3.2)
   c. Alkali- Alkali earth Metals (Table 3.3)
   d. Rare earth metals –Metalloids (Table 3.4)
2. Ternary metal chalcogenides
   a. Transition metal—Transition metal—Chalcogenide (Table 3.5)
   b. Transition metal—Post transition metal—Chalcogenide (Table 3.6)
   c. Alkali – Alkali earth metal – Transition metal Chalcogenide (Table 3.7)
   d. Transition metal—Metalloid—Chalcogenide (Table 3.8)
   e. Alkali, Alkali Earth Metal—Post transition metal—Chalcogenide (Table 3.9)
   f. Post transition metal—Metalloid—Chalcogenide (Table 3.10)
   g. Alkali, Alkali Earth Metal—Metalloid—Chalcogenide (Table 3.11)
3. Quaternary metal chalcogenides
   a. First, Second and Third Metal Chalcogenides (Table 3.12)
   The following section focuses on the latter.

3.1.3.1 *Binary metal chalcogenides*

Binary metal chalcogenides (BMCs) are a large group of materials having at least one metal element such as transition metals. BMSs have received great attention for their applications in thermo-, opto-, photo-, and piezo- electronics. These materials are widely used in sensors and biosensors via direct solution processing of these semiconductor thin films by roll-to-roll printing on flexible substrates. However, the low solubility of metals and chalcogenide in common solvents is one of the main drawbacks which cause many impurities in the final products. In a common method for preparation of binary metal chalcogenides, pure metal solutions of copper (Cu), zinc (Zn), Indium (In), etc. or binary metal chalcogenides (e.g., Cu_2S, Cu_2Se, CuS, CuSe, SnS, SnSe, In_2S_3, In_2Se_3, Ag_2S, and Ag_2Se) are prepared by dissolving bulk powders in solvent mixtures of primary amine [e.g., butylamine (BA), hexylamine (HA), etc.] and 1,2-ethanedithiol (EDT) at ambient conditions.

TABLE 3.1 Binary metal chalcogenide (transition metals) X = S, Se, Te (oxidation number -2).

1st row	**Sc**	**Ti**	**V**	**Cr**	**Mn**	**Fe**	**Co**	**Ni**	**Cu**	**Zn**
Oxidation numbers	+3	+4	+3, +4, + 5	+2, +3, + 6	+2, + 7	+2, + 3	+2, + 3	+2	+1, +2	+2
General formulas	Sc_2X_3	TiX_2	V_2X_3	CrX	MnX	FeX	CoX	NiX	Cu_2X, CuX	ZnX
Chemical formulas	Sc_2S_3, Sc_2Se_3, Sc_2Te_3	TiS_2, $TiSe_2$, $TiTe_2$	V_2S_3, V_2Se_3, V_2Te_3	CrS, Cr_2S_3, Cr_2Se_3,	MnS	FeS, Fe_2S_3, FeSe, Fe_2S_{e3}, Fe_2Se	CoS, CoSe, Co_2S_3, Co_2Se_3	NiS, NiSe, NiTe	Cu_2S, CuS Cu_2Se	ZnS, ZnSe, ZnTe
2nd row	Y	Zr	Nb	Mo	Tc	Ru	Rh	Pd	Ag	Cd
Oxidation numbers	+3	+4	+2, + 4	+4	+4	+4	+3	+2	+1	+2
General formulas	Y_2X_3	ZrX_2	NbX	MoX_2	TcX_2	RuX_2	Rh_2X_3	PdX	Ag_2X	CdX
Chemical formulas	Y_2S_3	ZrS_2	NbS, NbS_2	MoS_2, $MoSe_2$, $MoTe_2$	TcS_2	RuS_2, $RuSe_2$	Rh_2S_3	PdS, PdSe, PdTe	Ag_2S, Ag_2Se, Ag_2Te	CdS, CdSe, CdTe

TABLE 3.2 Binary metal chalcogenide (post transition metals).

Groups	IIIA	IVA	VA
Elements	Al, Ga, In, Tl	Sn, Pb	Bi
Oxidation numbers	+3	+2, + 4	+3
Chemical formulas	Al_2S_3, Al_2Se_3, GaS, Ga_2S_3, GaSe, In_2S_3, In_2Se_3, Tl_2S_3	SnS, SnS_2, PbS, PbSe	Bi_2S_3, Bi_2Se_3

TABLE 3.3 Binary metal chalcogenide (alkali—alkaline earth metals).

Groups	Alkali metals	Alkali earth metals
Elements	Li, Na, K, Rb, Cs	Be, Mg, Ca, Sr, Ba
Oxidation numbers	+1	+2
Chemical Formulas	Li_2S, Li_2Se, Li_2Te, Na_2S, Na_2Se, Na_2Te, K_2S, K_2Se, Rb_2S, Rb_2Se, Rb_2Te, Cs_2S, Cs_2Se	BeS, BeSe, MgS, MgSe, MgTe, CaS, CaSe, SrS, BaS, BaSe

TABLE 3.4 Binary metal chalcogenide (rare earth metals - metalloids).

Metalloids				Rare earth metals		
IVA	VA	IVA	IIIA			Groups
Te, Po	As, Sb,	Si, Ge	B	Yb	Ce	Elements
+4	+3, +5	+4	+3	+3	+3, +4	Oxidation numbers
TeS_2, $TeSe_2$	Sb_2S_3, Sb_2Se_3, Sb_2Te_3, As_2S_3, As_4S_5, As_2S_2, As_2Se_3	GeS_2, $GeSe_2$, $GeTe_2$, SiS_2, $SiSe_2$, $SiTe_2$	B_2S_3, B_2Se_3, B_2Te_3	Yb_2S_3, Yb_2Se_3, Yb_2Te_3	Ce_2S_3, CeS_2, Ce_2Se_3, $CeSe_2$	Chemical formulas

We classify the binary metal chalcogenides into four groups based on their elemental compositions: binary metal chalcogenides-based transition metals (Table 3.1), post-transition metals (Table 3.2), alkali—alkali earth metals (Table 3.3), and rare earth metals —metalloids (Table 3.4). Various oxidation states of 1st-row metals with chalcogens (S, Se, and Te) can form binary metal chalcogenides. Interestingly, the transition metals with multidisciplinary oxidation states can form different metal chalcogenides, like vanadium (V) ions with III, IV, and V oxidation or

TABLE 3.5 Ternary metal chalcogenide (transition metal–transition metal–chalcogenide) bimetallic chalcogenide (transition metal–transition metal) X = S, Se, Te.

Molar composition	Transition metal chalcogenide							Transition metal Chalcogenide
	Cr_2X_3	MnX	FeX	Co_2X_3	CoX	Ag_2X	CdX	
1:1	Cr_2NiS_4 [$NiCr_2S_4$]	$MnNiS_2$ [$NiMnS_2$]	$FeNiS_2$	Co_2NiS_4 [$NiCo_2S_4$]	$CoNiS_2$ [$NiCoS_2$]	Ag_2NiS_2		NiX
1:2			$FeNi_2S_3$			$Ag_2Ni_2S_4$		
2:1			Fe_2NiS_3 [$NiFe_2S_3$]		Co_2NiS_3 [$NiCo_2S_3$]			
1:1	$CrCuS_2$ [$CuCrS_2$]	$MnCu_2S_2$ [Cu_2MnS_2]	$FeCu_2S_3$ [Cu_2FeS_3]	$CoCuS_2$	$CoCu_2S_2$ [Cu_2CoS_2]	AgCuS		Cu_2X
3:1						Ag_3CuS_2		
2:1			$Fe_2Cu_2S_3$ [$Cu_2Fe_2S_3$]					
1:1	Cr_2CuS_4 [$CuCr_2S_4$] Cr_2CuSe_4 $CuCr_2Se_4$	$MnCuS_2$ [$CuMnS_2$], $MnCuSe_2$ [$CuMnSe_2$]	$FeCuS_2$ [$CuFeS_2$], $FeCuSe_2$ [$CuFeSe_2$]	Co_2CuS_4	$CoCuSe_2$	Ag_2CuS_2	$CuCdS_2$ [$CuCdS_2$]	CuX
1:2						$Ag_2Cu_2S_3$		
2:1					Co_2CuSe_3 [Co_2CuSe_3]			

(Continued)

TABLE 3.5 (Continued)

Molar composition	Transition metal chalcogenide							Transition metal Chalcogenide
	Cr_2X_3	MnX	FeX	Co_2X_3	CoX	Ag_2X	CdX	
1:1	Cr_2ZnS_4 [$ZnCr_2S_4$]	$MnZnS_2$, $MnZnSe_2$	$FeZnS_2$	Co_2ZnS_4 [$ZnCo_2S_4$]	$CoZnS_2$ [$ZnCoS_2$]		$CdZnS_2$, $CdZnSe_2$	ZnX
1:2		$MnZn_2S_3$ [Zn_2MnS_3]	$FeZn_2S_3$ [Zn_2FeS_3]					
2:3			$Fe_2Zn_3S_5$					
1:1	$MoCr_2S_6$	$MnMoS_4$	$FeMoS_4$		$CoMoS_4$	Ag_2MoS_4	$CdMoS_4$	MoX_3
2:1			Fe_2MoS_5 [$MoFe_2S_5$]			Ag_4MoS_5		
3:2			$Fe_3Mo_2S_9$					
1:1		$MnMoS_3$	$FeMoS_3$		$CoMoS_3$	Ag_2MoS_3	$CdMoS_3$ [$MoCdS_3$]	MoX_2
2:1					Co_2MoS_4	Ag_4MoS_4		
1:1	$MoCrS_3$	$MnMo_2S_4$	$FeMo_2S_4$	$CoMoS_3$	$CoMo_2S_4$	$AgMoS_2$		Mo_2X_3
1:1		$MnFe_2S_4$	Fe_3S_4		$CoFe_2S_4$ $CoFe_2Se_4$	$AgFeS_2$		Fe_2X_3

TABLE 3.6 Ternary metal chalcogenide (transition metal–post transition metal–chalcogenide) bimetallic chalcogenide (transition metal–post transition metal) (X = S, Se, Te).

Molar composition	Transition metal chalcogenide				Post transition metal chalcogenide
	MnX	FeX	CoX	Cu_2X	
1:1	$MnAl_2S_4$	$FeAl_2S_4$	$CoAl_2S_4$	$CuAlS_2$	Al_2X_3
2:1	$Mn_2Al_2Se_5$				
1:1	$MnIn_2Se_4$	$FeIn_2Se_4$	$CoIn_2S_4$	$CuInS_2$	In_2X_3
2:1			$Co_2In_2Se_5$		
1:2				$Cu_2In_4S_7$	
1:1	$MnSnS_3$	$FeSnS_3$	$CoSnS_3$	Cu_2SnSe_3	SnX_2
2:1	Mn_2SnS_4	Fe_2SnS_4	Co_2SnS_4	Cu_4SnS_4	
1:2				$Cu_2Sn_2S_5$	
1:1	$MnBi_2Te_4$	$FeBi_2Te_4$	$CoBi_2Te_4$	$Cu_2Bi_2S_4$	Bi_2X_3
2:1	$Mn_2Bi_2Te_5$		$Co_2Bi_2Te_5$		
3:1				Cu_3BiS_3	

TABLE 3.7 Ternary metal chalcogenide (alkali, alkali earth metal–transition metal chalcogenide) (X = S, Se, Te).

Molar composition	Alkali, alkali earth metal chalcogenide				Transition metal chalcogenide
	Na_2X	K_2X	SrX	BaX	
1:1	Na_2FeS_2		$SrFeS_2$	$BaFeTe_2$	FeX
2:1		K_4FeS_3		Ba_2FeSe_3	
1:2	$Na_2Fe_2S_3$		$SrFe_2Se_3$	$BaFe_2S_3$	
1:1		K_2CoSe_2		$BaCoS_2$	CoX
2:1	Na_4CoX_3			Ba_2CoS_3	
1:2			$SrCo_2X_3$		
1:1		K_2ZnS_2		$BaZnX_2$	ZnX
2:1	Na_4ZnSe_3		Sr_2ZnS_3		
1:3		$K_2Zn_3S_4$		$BaZn_3X_4$	
1:1	$Na_2Fe_2Se_4$		$SrFe_2S_4$	$BaFe_2Se_4$	Fe_2X_3
2:1		$K_4Fe_2X_5$			
1:2	$Na_2Fe_4X_7$		$SrFe_4X_7$	$BaFe_4X_7$	

TABLE 3.8 Ternary metal chalcogenide (transition metal–metalloid–chalcogenide) bimetallic chalcogenide (transition metal–metalloid) X = S, Se, Te.

Molar composition	Transition metal chalcogenide						Metalloid–chalcogenide	
	MnX MnS, MnSe, MnTe	**FeX FeS, FeSe, FeTe**	**CoX CoS, CoSe**	**NiX NiS, NiSe**	**CuX, Cu_2X CuS, Cu_2S**	**ZnX ZnS, ZnSe**	**Chemical formulas**	**General formula**
2:1	Mn_2GeS_4, Mn_2GeSe_4, Mn_2GeTe_4	Fe_2GeS_4, Fe_2GeSe_4, Fe_2GeTe_4	Co_2GeS_4, Co_2GeSe_4,	Ni_2GeS_4, Ni_2GeSe_4	Cu_2GeS_4	Zn_2GeS_4, Zn_2GeSe_4	GeS_2, $GeSe_2$, $GeTe_2$	GeX_2
1:1	$MnGeS_3$, $MnGeSe_3$, $MnGeTe_3$	$FeGeS_3$	$CoGeSe_3$	$NiGeSe_3$	$CuGeS_3$, Cu_2GeS_3	$ZnGeS_3$, $ZnGeSe_3$		
1:2					$CuGe_2S_5$			
2:1	Mn_2SiS_4, Mn_2SiSe_4, Mn_2SiTe_4, Mn_2SiS_3Se	Fe_2SiS_4, Fe_2SiSe_4	Co_2SiS_4,	Ni_2SiS_4, Ni_2SiTe_4			SiS_2, $SiSe_2$, $SiTe_2$	SiX_2
1:1					Cu_2SiS_3, Cu_2SiSe_3, Cu_2SiTe_3			
2:1	$Mn_2Sb_2S_5$, $Mn_2Sb_2Te_5$	$Fe_2Sb_2Te_5$			$Cu_2Sb_2S_5$, $Cu_2Sb_2Se_5$		Sb_2S_3, Sb_2Se_3, Sb_2Te_3	Sb_2X_3
1:1	$MnSb_2S_4$, $MnSb_2Se_4$, $MnSb_2Te_4$	$FeSb_2S_4$, $FeSb_2Se_4$	$CoSb_2S_4$	$NiSb_2Se_4$, $NiSb_2Te_4$	$CuSb_2S_4$	$ZnSb_2S_4$		
2:1			$Co_2As_2S_5$, $Co_2As_2Se_5$	$Ni_2As_2S_5$	$Cu_2As_2S_5$, $Cu_2As_2Se_5$	$Zn_2As_2S_5$	As_2S_3, As_4S_5, As_2S_2, As_2Se_3	As_2X_3
1:1		$FeAs_2S_4$	$CoAs_2S_4$, $CoAs_2Se_4$	$NiAs_2S_4$	$CuAs_2S_4$			
					Cu_3AsS_4, Cu_3AsSe_4			

TABLE 3.9 Ternary metal chalcogenide (alkali, alkali earth metal—post transition metal—chalcogenide) bimetallic chalcogenide (alkali, alkali earth metal—post transition metal) X = S, Se, Te.

Molar composition	Alkali, alkali earth metal chalcogenide										Post transition metal Cha lcogenide
	Li_2S, Li_2Se, Li_2Te	Na_2S, Na_2Se, Na_2Te	K_2S, K_2Se	Rb_2S, Rb_2Se, Rb_2Te	Cs_2S, Cs_2Se	BeS	MgS, MgSe	CaS, CaSe, CaTe	SrS, SrSe, SrTe	BaS, BaSe, BaTe	
1:1	$LiAlS_2$, $LiAlSe_2$, $LiAlTe_2$	$NaAlS_2$, $NaAlSe_2$, $NaAlTe_2$	$KAlS_2$, $KAlSe_2$, $KAlTe_2$	$RbAlS_2$, $RbAlSe_2$, $RbAlTe_2$	$CsAlS_2$, $CsAlSe_2$, $CsAlTe_2$	$BeAl_2S_4$	$MgAl_2S_4$, $MgAl_2Se_4$, $MgAl_2Te_4$	$CaAl_2S_4$, $CaAl_2Se_4$, $CaAl_2Te_4$	$SrAl_2S_4$, $SrAl_2Se_4$, $SrAl_2Te_4$	$BaAl_2S_4$, $BaAl_2Se_4$, $BaAl_2Te_4$	Al_2X_3 (Al_2S_3, Al_2Se_3)
2:1		$Na_4Al_2S_5$		$Rb_4Al_2S_5$			$Mg_2Al_2S_5$, $Mg_2Al_2Se_5$	$Ca_2Al_2S_5$	$Sr_2Al_2S_5$	$Ba_2Al_2S_5$	
1:1	$LiInS_2$ $LiInSe_2$ $LiInTe_2$	$NaInS_2$ $NaInSe_2$ $NaInTe_2$	$KInS_2$ $KInSe_2$ $KInTe_2$	$RbInS_2$ $RbInSe_2$ $RbInTe_2$	$CsInS_2$ $CsInSe_2$ $CsInTe_2$	$BeIn_2S_4$	$MgIn_2S_4$ $MgIn_2Se_4$ $MgIn_2Te_4$	$CaIn_2S_4$ $CaIn_2Se_4$ $CaIn_2Te_4$	$SrIn_2S_4$, $SrIn_2Se_4$, $SrIn_2Te_4$	$BaIn_2S_4$, $BaIn_2Se_4$, $BaIn_2Te_4$	In_2X_3 (In_2S_3 In_2Se_3, In_2Te_3)
2:1			$K_4In_2S_5$	$Rb_4In_2S_5$	$Cs_4In_2S_5$					$Ba_2In_2S_5$, $Ba_2In_2Se_5$	
1:1	Li_2SnS_3, Li_2SnSe_3	Na_2SnS_3, Na_2SnSe_3, Na_2SnTe_3	K_2SnS_3, K_2SnSe_3, K_2SnTe_3	Rb_2SnS_3, Rb_2SnSe_3,		$BeSnS_2$		$CaSnS_2$	$SrSnS_2$, $SrSnSe_2$, $SrSnS_3$, $SrSnSe_3$	$BaSnS_2$, $BaSnSe_2$	SnX_2 (SnS, SnS_2)
2:1		Na_4SnS_4, Na_4SnSe_4, Na_4SnTe_4	K_4SnS_4, K_4SnSe_4, K_4SnTe_4					Ca_2SnS_4	Sr_2SnS_4	Ba_2SnS_4	
1:2			$K_2Sn_2S_5$, $K_2Sn_2Se_5$, $K_2Sn_2Te_5$	$Rb_2Sn_2S_5$, $Rb_2Sn_2Se_5$				$CaSn_2S_5$		$BaSn_2S_5$, $BaSn_2Se_5$,	

(*Continued*)

TABLE 3.9 (Continued)

Molar composition	Alkali, alkali earth metal chalcogenide										Post transition metal Cha lcogenide
	Li_2S, Li_2Se, Li_2Te	**Na_2S, Na_2Se, Na_2Te**	**K_2S, K_2Se**	**Rb_2S, Rb_2Se, Rb_2Te**	**Cs_2S, Cs_2Se**	**BeS**	**MgS, MgSe**	**CaS, CaSe, CaTe**	**SrS, SrSe, SrTe**	**BaS, BaSe, BaTe**	
1:1	$LiBiS_2$, $LiBiSe_2$, $LiBiTe_2$	$NaBiS_2$, $NaBiSe_2$, $NaBiTe_2$	$KBiS_2$, $KBiSe_2$, $KBiTe_2$	$RbBiS_2$, $RbBiSe_2$, $RbBiTe_2$	$CsBiS_2$, $CsBiSe_2$, $CsBiTe_2$		$MgBi_2S_4$	$CaBi_2S_4$, $CaBi_2Se_4$, $CaBi_2Te_4$	$SrBi_2S_4$, $SrBi_2Se_4$	$BaBi_2S_4$, $BaBi_2Se_4$	Bi_2X_3 (Bi_2S_3, Bi_2Se_3)
1:4			$K_2Bi_8Se_{13}$	$Rb_2Bi_8Se_{13}$	$Cs_2Bi_8Se_{13}$						
1:1		Na_2PbS_2, Na_2PbSe_2	K_2PbS_2, K_2PbSe_2,						$SrPbS_2$, $SrPbSe_2$	$BaPbS_2$, $BaPbSe_2$	PbS, PbSe

TABLE 3.10 Ternary metal chalcogenide (post transition metal–metalloid–chalcogenide) bimetallic chalcogenide (Post transition metal –Metalloid) X = S, Se, Te.

	Post transition metal chalcogenide					Metalloid–chalcogenide	
Molar composition	**Ga_2X_3 Ga $_2S_3$, Ga $_2Se_3$**	**In_2X_3 In_2S_3, In_2Se_3, In_2Te_3**	**SnX_2 SnS, SnS_2**	**Bi_2X_3 Bi_2S_3, Bi_2Se_3**	**PbX PbS, PbSe**	**Chemical formulas**	**General formula**
1:1	Ga_2GeS_5, Ga_2GeSe_5	In_2GeS_5, In_2GeSe_5, In_2GeTe_5	$SnGeS_3$, $SnGeSe_3$,		$PbGeS_3$, $PbGeSe_3$	GeS_2, $GeSe_2$, $GeTe_2$	GeX_2
1:2	$Ga_2Ge_2S_7$	$In_2Ge_2S_7$					
2:1			Sn_2SiS_4, Sn_2SiSe_4		Pb_2SiS_4, Pb_2SiSe_4	SiS_2, $SiSe_2$, $SiTe_2$	SiX_2
1:1			$SnSiS_3$, $SnSiSe_3$, $SnSiTe_3$		$PbSiS_3$, $PbSiSe_3$		
2:1			$Sn_2Sb_2S_5$	$Bi_4Sb_2S_9$	$Pb_2Sb_2S_5$, $Pb_2Sb_2Se_5$, $Pb_2Sb_2Te_5$	Sb_2S_3, Sb_2Se_3, Sb_2Te_3	Sb_2X_3
1:1		$InSbS_3$, $InSbSe_3$, $InSbTe_3$	$SnSb_2S_4$, $SnSb_2Se_4$, $SnSb_2Te_4$	$BiSbS_3$, $BiSbSe_3$, $BiSbTe_3$	$PbSb_2S_4$, $PbSb_2Se_4$, $PbSb_2Te_4$		
2:1					$Pb_2As_2S_5$ (Dufrenoysite), $Pb_2As_2Se_5$, $Pb_2As_2Te_5$	As_2S_3, As_4S_5, As_2S_2, As_2Se_3	As_2X_3
1:1		$InAsS_3$,	$SnAs_2S_4$, $SnAs_2Se_4$, $SnAs_2Te_4$	$BiAsS_3$	$PbAs_2S_4$, $PbAs_2Se_4$, $PbAs_2Te_4$		

TABLE 3.11 Ternary metal chalcogenide (alkali, alkali earth metal–metalloid–chalcogenide) bimetallic chalcogenide (alkali, alkali earth metal–metalloid) X = S, Se, Te.

Molar composition	Alkali, alkali earth metal chalcogenide										Metalloid–chalcogenide formulas
	Li_2S, Li_2Se, Li_2Te	Na_2S, Na_2Se, Na_2Te	K_2S, K_2Se	Rb_2S, Rb_2Se, Rb_2Te	Cs_2S, Cs_2Se	BeS	MgS, MgSe	CaS, CaSe, CaTe	SrS, SrSe, SrTe	BaS, BaSe, BaTe	
1:1	Li_2GeS_3, Li_2GeSe_3	Na_2GeS_3, Na_2GeSe_3	K_2GeS_3	Rb_2GeS_3, Rb_2GeSe_3, Rb_2GeTe_3	Cs_2GeS_3, Cs_2GeSe_3, Cs_2GeTe_3	—	$MgGeS_3$	—	$SrGeS_3$, $SrGeSe_3$	$BaGeS_3$, $BaGeSe_3$,	GeS_2, $GeSe_2$, $GeTe_2$
2:1	Li_4GeS_4, Li_4GeSe_4	Na_4GeS_4, Na_4GeSe_4, Na_4GeTe_4	K_4GeS_4, K_4GeSe_4	Rb_4GeS_4	Cs_4GeS_4	—	Mg_2GeS_4, Mg_2GeSe_4, Mg_2GeTe_4	Ca_2GeS_4, Ca_2GeSe_4, Ca_2GeTe_4	Sr_2GeS_4, Sr_2GeSe_4, Sr_2GeTe_4	Ba_2GeS_4, Ba_2GeSe_4, Ba_2GeTe_4	
1:1	Li_2SiS_3, Li_2SiSe_3	Na_2SiS_3, Na_2SiSe_3, Na_2SiTe_3	K_2SiS_3, K_2SiSe_3			$BeSiS_3$,	$MgSiS_3$	$CaSiS_3$			SiS_2, $SiSe_2$, $SiTe_2$
1:2		$Na_2Si_2S_5$, $Na_2Si_2Se_5$			$Cs_2Si_2S_5$			$CaSi_2S_5$	$SrSi_2S_5$	BaSi2S_5	
1:1		$Na_2Sb_2S_4$, $Na_2Sb_2Se_4$	$K_2Sb_2S_4$	$Rb_2Sb_2S_4$	$Cs_2Sb_2S_4$			$CaSb_2S_4$	$SrSb_2S_4$	$BaSb_2S_4$, $BaSb_2Se_4$	Sb_2S_3, Sb_2Se_3, Sb_2Te_3
3:2	$Li_6Sb_4S_9$	$Na_6Sb_4S_9$							$Sr_3Sb_4S_9$, $Sr_3Sb_4Se_9$		
3:1	Li_3SbS_3, Li_3SbSe_3	Na_3SbS_3, Na_3SbSe_3, Na_3SbTe_3	K_3SbS_3, K_3SbSe_3, K_3SbTe_3	Rb_3SbS_3, Rb_3SbSe_3, Rb_3SbTe_3	Cs_3SbS_3, Cs_3SbSe_3,	—			$Sr_3Sb_2S_6$		
1:1		$Na_2As_2S_4$	$K_2As_2S_4$					$CaAs_2S_4$	$SrAs_2S_4$	$BaAs_2S_4$	As_2S_3, As_4S_5, As_2S_2, As_2Se_3
1:2		$Na_2As_4S_7$	$K_2As_4S_7$								

TABLE 3.12 Quaternary metal chalcogenides classifications-based molar composition.

1st	2nd							3rd
	FeX	CoX	ZnX	CdX	Al_2X_3	Ga_2X_3	In_2X_3	
Cu_2X	1:1:1 Cu_2FeSnS_4 $Cu_2FeSnSe_4$	1:1:1 Cu_2CoSnS_4 $Cu_2CoSnSe_4$	1:1:1 $Cu_2ZnSnSe_4$	1:1:1 Cu_2CdSnS_4 $Cu_2CdSnSe_4$	1:1:2 $CuAlSnSe_4$	1:1:2 $CuGaSnSe_4$	1:1:2 $CuInSnSe_4$	SnX_2
	1:1:1 Cu_2FeGeS_4 $Cu_2FeGeSe_4$ $Cu_2FeGeTe_4$	1:1:1 Cu_2CoGeS_4 $Cu_2CoGeSe_4$	1:1:1 Cu_2ZnGeS_4 $Cu_2ZnGeSe_4$	1:1:1 Cu_2CdGeS_4 $Cu_2CdGeSe_4$	1:1:2 $CuAlGeSe_4$	1:1:2 $CuGaGeSe_4$	1:1:2 $CuInGeSe_4$	GeX_2
	1:2:3 $CuFeIn_3S_6$		1:2:3 $CuZnIn_3S_6$			1:2:3 $CuGa_2In_3S_8$	—	In_2X_3
	1:1:1 $Cu_2FeSiSe_4$	1:1:1 Cu_2CoSiS_4	1:1:1 Cu_2ZnSiS_4 $Cu_2ZnSiSe_4$	1:1:1 Cu_2CdSiS_4 $Cu_2CdSiSe_4$	1:1:2 $CuAlSiS_4$ $CuAlSiSe_4$	1:1:2 $CuGaSiS_4$ $CuGaSiSe_4$		SiX_2
Ag_2X	1:1:1 Ag_2FeSnS_4 $Ag_2FeSnSe_4$		1:1:1 Ag_2ZnSnS_4 $Ag_2ZnSnSe_4$	1:1:1 Ag_2CdSnS_4 $Ag_2CdSnSe_4$ $Ag_2CdSnTe_4$	1:1:2 $AgAlSnSe_4$	1:1:2 $AgGaSnS_4$ $AgGaSnSe_4$	1:1:2 $AgInSnSe_4$	SnX_2
	1:1:1 Ag_2FeGeS_4 $Ag_2FeGeSe_4$ $Ag_2FeGeTe_4$	1:1:1 Ag_2CoGeS_4 $Ag_2CoGeSe_4$	1:1:1 Ag_2ZnGeS_4 $Ag_2ZnGeSe_4$	1:1:1 Ag_2CdGeS_4 $Ag_2CdGeSe_4$	1:1:2 $AgAlGeSe_4$	1:1:2 $AgGaGeSe_4$	1:1:2 $AgInGeSe_4$	GeX_2
			1:1:14 $Ag Zn_7InS_9$			1:2:3 $AgGa_2In_3S_8$	—	In_2X_3

(Continued)

TABLE 3.12 (Continued)

1st	2nd							3rd
	FeX	CoX	ZnX	CdX	Al_2X_3	Ga_2X_3	In_2X_3	
			[$AgInZn_7S_9$]					
	1:1:1 Ag_2FeSiS_4 $Ag_2FeSiSe_4$		1:1:1 Ag_2ZnSiS_4 $Ag_2ZnSiSe_4$	1:1:1 Ag_2CdSiS_4 $Ag_2CdSiSe_4$			1:1:1 $Ag_2In_2SiS_6$ $Ag_2In_2SiSe_6$	SiX_2
Li_2X	1:1:1 Li_2FeSnS_4						1:1:2$LiInSnS_4$	SnX_2
	1:1:1 Li_2FeGeS_4			1:1:1 Li_2CdGeS_4 $Li_2CdGeSe_4$		1:1:2 Li $GaGeS_4$		GeX_2
							—	In_2X_3
	1:1:1 Li_2FeSiS_4		1:1:1 Li_2ZnSiS_4				1:1:1 $Li_2In_2SiS_6$ $Li_2In_2SiSe_6$	SiX_2

chromium (Cr) ions with II, III, and VI oxidation states (Table 3.1). Moreover, chalcogens can bond with 2nd-row transition metals like yttrium (Y^{3+}), zirconium (Zr^{4+}), niobium ($Nb^{2+,4+}$), molybdenum (Mo^{4+}), etc.

The post-transition metals include any metal in groups 3, 4, and 5 of the periodic table of elements. Seven elements of post-transition metals are aluminum (Al), gallium (Ga), indium (in), tin (Sn), thallium (Tl), lead (Pb), and bismuth (Bi). The chemistry of post transition metals is dominated by the group oxidation state N and a lower N-2 oxidation state, which is associated with occupation of a metal s^2 lone pair. These elements are capable to react with chalcogens to form metal chalcogenides. Table 3.2 compares various chemical formulas of metal chalcogenides based on post-transition metals.

Alkali metals are highly reactive elements in Group 1 of the periodic table including lithium (Li), sodium (Na), potassium (K), rubidium (Rb), cesium (Cs), and francium (Fr). The high reactivity of these elements can be related to their low ionization energies and larger atomic radii. The alkali metals can easily react with the elements of Group 6, that is, chalcogens to form metal chalcogenides (MCs), as; $2A_{(s)} + B_{(s)} \rightarrow A_2B_{(s)}$; where A and B are alkali metals and chalcogens, respectively. Table 3.3 represents various possibilities of Alkali metals-based chalcogenides.

The second most reactive element after alkali metals are alkaline earth metals (Group 2). In this group of elements, there is beryllium (Be), magnesium (Mg), calcium (Ca), strontium (Sr), barium (Ba), and radium (Ra). These elements possess low first and second ionization energies, therefore group 2 elements can form ionic compounds. Among the elements in this group, Be is the lightest element with superior ionization energy, hence "Be" can form largely covalent compounds. The reactivity of the alkaline earth metals against chalcogens (Group 6) is almost similar to the alkali metals. The alkaline earth metals and chalcogens can form binary chalcogenides in a 1:1 ratio (Table 3.3). However, in a lower ratio of the elements (alkaline earth metal: chalcogen), the salts containing polychalcogenide ions can be formed.

The rare earth (RE) elements consist of 17 elements consisting of 15 lanthanides plus scandium (Sc) and yttrium (Y). RE is not essential to plants. The magnetic and luminescent performances of REs are superior as compared to other elements, therefore used widely in energy storage, energy conversion, and smart technologies.

REs can react with chalcogens (Group 6) with a variety of stoichiometries of REs and chalcogens to form binary metal chalcogenides. The use of binary REs-based transition metal chalcogenides is widely used as a catalyst. These classes of materials have been investigated effectively in energy and sensing devices, both theoretical and experimental. However, the activity of these binary chalcogenides is raised just from the edge sites, whereas the basal planes are catalytically inactive. This limitation has been overcome by maximizing the number of edge sites of crystalline and amorphous phases. Some important examples of binary RE chalcogenides are summarized in Table 3.4. On the periodic table, metalloids include boron (B), silicon (Si), germanium (Ge), arsenic (As), antimony (Sb), and tellurium (Te). These elements lay in groups 3–6 of periodic tables on either side of the dividing line between metals and nonmetals. Metalloids with chalcogens generate compounds with different properties than pristine metalloids; more reactive and more toxic. Metalloid chalcogenides (Me_2Ch_3) mostly exist in nature as a crystalline minerals. Metalloids are used for their +3, +4, and +5 oxidations in combination with chalcogenides (Table 3.4).

3.1.3.2 *Ternary metal chalcogenides*

Ternary metal chalcogenides have been used as phase change and photovoltaic materials. These materials can be subject to phase-change over operating temperature (i.e., exist in a crystalline state or an amorphous state). The atomic arrangements of phase-change materials are more ordered, with lower electrical resistance, in their crystalline state as compared to the amorphous state. Ternary metal chalcogenides can be classified based on their elemental compositions as:

- Bimetallic chalcogenide-based double Transition metals
- Bimetallic chalcogenide-based Transition metal and Post transition metal
- Bimetallic chalcogenide-based Alkali or alkali earth metal, and Transition metal
- Bimetallic chalcogenide-based Transition metal and Metalloid

In bimetallic chalcogenide-based transition metals (Table 3.5), the final products (i.e., ternary metal chalcogenide) are composed of two transition ions which are generally in the form of $M_{tr.x}$ - $M_{tr.y}$ - Ch_z with different molar ratios of transition metals such as 1:1, 1:2, 2:1, 1:1, 3:1, 2:1, 2:3, and 3:2. In this category, the primary transition metals are Mn, Cr, Fe, Co, Ag, Cd, while the secondary transition metals can be Ni, Cu, Zn, Mo, and Fe. These materials are generally antiferromagnetic semiconductors, which are classified under various phase prototype indices. For instance, the phase prototype of $NiCr_2S_4$ is Mo_2CoS_4, or the phase prototype of $FeNiS_2$ is $CuFeS_2$. The phase prototype is an index of the various crystal structures by the prototype compound. The table of prototype index including Wyckoff positions are added as Index A. Wyckoff position is a point belonging to a set of points for which site symmetry groups are conjugate subgroups of the space group, or in the other words those positions which the structure occupies in its space group.

Ternary Metal Chalcogenide-based transition metal, a post-transition metal, and chalcogens are also designed for various catalytic and optical applications. For example, $MnAl_2S_4$ (optical energy gap 3.75 eV) and $MnAl_2Se_4$ (optical energy gap 3.21 eV) single crystals are used for optical applications, where synthesized by chemical transport reaction method. Doublet emission peaks are growth at 450 and 603 nm in the $MnAl_2S_4$, while shifting into the higher values (488 and 655 nm) in the $MnAl_2Se_4$ single crystal. In this class of materials, the composition of transition metals-based Mn, Fe, Co, and Cu and post-transition metals such as Al, In, Sn, and Bi, with S, Se, and Te are reported (Table 3.6).

Group I and II metals reacting with transition metals, in the presence of chalcogens can also form ternary metal chalcogenides such as Na_2FeS_2, $SrFe_2Se_3$, etc. The molar ratio of alkali (alkali metal) to transition metals varies from 1:1, 2:1, 1:3, and 1:2 (Table 3.7). In transition metal sites, Iron (Fe), Cobalt (Co), and Zinc (Zn) are prominent, while in alkali (or alkali metal) sites, sodium (Na), potassium (K), strontium (Sr), and barium (Ba) are significant. This class of bimetallic chalcogenides is rarely investigated, most probably due to the difficulties in their synthesis and narrow range of applications. Let's have a look at the structural features of K_2CoSe_2. This chalcogenide crystallizes in the orthorhombic Ibam space group in three-dimensional. In this structure, K^{+1} bonds in a six-coordinate geometry into six equivalent Se^{2-} atoms. The final magnetic momentum of this ferromagnetic 3D structure is 3.000 μB, with a density of 3.36 g/cm^3. K_2CoSe_2 can decompose into the KSe + Co_9Se_8 + K_2Se.

Bimetallic chalcogenides, based on transition metal, and metalloid, are well-known groups of ternary metal chalcogenides (Table 3.8). These materials are formed from three structural compartments namely transition metals-chalcogen such as MnX, FeX, CoX, NiX, CuX, ZnX (where X is S, Se, Te), metalloid-chalcogen like GeX_2, SiX_2, Sb_2X_3, and As_2X_3. The coordination between the above-mentioned materials can form a complex chalcogenide structure. For instance, 2MnS + GeS_2 - Mn_2GeS_4, or 2CoS + SiS_2 - Co_2SiS_4. The molar compositions are varied between 2 and 1 of each primary metal compartment. Some complex chalcogenides in this class contain two chalcogens instead of one such as Mn_2SiS_3Se. Magnetic excitations from frustrated lattices of magnetic atoms in Mn_2SiS_3Se are important in quantum studies. This material has a spinel-like structure consisting of one quadrivalent and two divalent cations in the form of $A_2^{2+}B^{4+}X_4$ where the A sites consist of the two crystallographically independent sites of 4a and 4c. From these compositions, the final products can be $Mn_2SiS_{4-x}Se_x$—Mn_2SiS_4 and Mn_2SiSe_4. These materials are antiferromagnetic below their Neel temperature.

3.1.3.3 *Quaternary metal chalcogenides*

In last two decades, quaternary chalcogenides have attracted attention as low-cost alternatives to conventional photovoltaic materials. These materials can be synthesized via three different wet organometallic routes; however, the controlled morphology of quaternary nanocrystals is still a challenge. A quaternary metal chalcogenide forms from stoichiometric combinations of three metal chalcogenides in a single or double step reaction. Whenever oxidation numbers of the metal chalcogenides change, likewise the molar composition of the three reactants alters. Quaternary chalcogenides are widely used in photovoltaic systems. These materials combine many advantageous elemental compositions and are technically characterized for optoelectronic applications. These materials are formed from their respective ternary metals pus chalcogens. Appropriate band gap, high optical absorption coefficient, low toxicity, and abundance are some important properties of quaternary metal chalcogenides. Based on the metal compositions, these complex materials can be classified as:

- Ternary Metal Chalcogenide (Alkali, Alkali Earth Metal—Post transition metal—Chalcogenide (Table 3.9)
- Ternary Metal Chalcogenide (Post transition metal—Metalloid—Chalcogenide) (Table 3.10)
- Ternary Metal Chalcogenide (Alkali, Alkali Earth Metal – Metalloid – Chalcogenide) (Table 3.11)
- Combinations of three metal chalcogenides (Table 3.12)

Quaternary metal chalcogenides are the last group of chalcogenides formed from stoichiometric combinations of three metal chalcogenides in single or double-step reactions. Whenever oxidation numbers of the metal chalcogenides change, likewise the molar composition of the three reactants alters. For example:

$$Cu_2X + FeX + SnX_2 = 2CuX + FeX + SnX = Cu_2FeSnX_4 (X = S, Se, Te)$$

This group of metal chalcogenides shows high flexibility for tuning the band gap. This adjustment is without clinging to toxic elements which are generally used for tunning the band gap. These compounds are used as p-type semiconducting materials for light harvesting in thin film solar cells and sensing devices. The light-harvesting is obtained via optimization of the band gap and energy levels by controlling composition and deposition conditions. In one classification, the 1st element commonly from main group elements (Li^+, Cu^+, Ag^+) anchors with 2nd elements from 1st-row transition metal ions (Fe^{2+}, Co^{2+}) or post-transition metal ions (Zn^{2+}, Cd^{2+}, Al^{3+}, Ga^{3+}, In^{3+}), where finally join with 3rd elements Si^{4+}, Ge^{4+}, Sn^{4+}, and In^{3+} (Table 3.12).

The combination of two group VI elements or two groups III elements can form "pseudo-quaternary" systems.

In summary, quaternary semiconductor-based metal chalcogenides provide some advantages over bulk materials such as:

- The flexibility of the band gap and electronic energy levels,
- Controllability of their composition and internal structure,
- Excellent physical and chemical features,
- Easy synthesis and processability,
- Nontoxic and economically feasible production.

3.1.4 Physical properties, optical and thermo-optic properties

Chalcogenide glasses are convenient materials for solid membranes of various types of solid-state chemical sensors [conventional ion-selective electrodes (ISE), ion-selective field-effect transistors (ISFET), miniature silicon-based sensors], which can be developed both as selective discrete sensors and low-selective sensor arrays [13].

Although it was only recently that anisotropic metal chalcogenide (MC) nanocrystals have been investigated with much interest (with regards to their applications), remarkable advances have been made with anisotropic nanocrystals in diverse areas of modern technology. These MC nanocrystals are also strong candidate materials for thermoelectric devices [13].

Energy harvested from heat loss and cooling electronic devices by effective heat transfer has motivated the studies on chalcogenide-based thermoelectric devices. In photovoltaic and photodetector devices, MC nanocrystals have been shown to absorb sunlight, hence they can play as exciton generators and charge transport materials [14].

The MC nanocrystals with a layered crystal structure and their composites with carbon materials are promising candidates that can intercalate Li^+ ions reversibly, which can be used as electrodes for batteries [15]. In addition to these applications, new potential novel applications are recently being investigated for these MCs. As an alternative to noble metals, doped MC nanostructure has attracted interest as an efficient catalyst for the oxygen reduction reaction [16], and as a patternable material whose localized surface plasmon resonance is tunable [17,18]. The topological insulating characteristics and superconductivity of MC nanostructures are also considered new areas of research for this class of materials [19,20]. Anisotropic MC nanocrystals have predominantly been synthesized in solutions because the size and shape of the nanocrystals can be controlled precisely, and solution-processed printing is the desired inexpensive way to produce MC films. The well-known

Gibbs–Wulff determines the shape of the nanocrystal in a way that the total surface energy of the crystal faucets minimizes.

Soft chalcogens (S, Se, Te) and their respective metal chalcogens have the preferred directional growth, which means the existence of thermodynamically stable crystal growth. Se and Te are well known to grow into one-dimensional nanostructures due to their strong bond along the *c*-axis. 2D metal chalcogens such as Bi_2Se_3 and Bi_2Te_3 can grow in a thermodynamic preference without any shape-guiding agents. It varies in the metal species and the stoichiometric ratio of the metal to chalcogen (A_xB_y, A = metal, B = S, Se, Te).

It must be mentioned that confinement effects lead to some of the most fundamental manifestations of nanoscale phenomena in chalcogens [21]. Novel optical properties appear in nanoscales chalcogenides as a result of such effects and are being exploited for information, biological, sensing, and energy technologies. The optical properties of chalcogen nanoparticles arise from a complex electrodynamic effect that is strongly influenced by the surrounding dielectric medium. Like metal oxides, light impinging on metallic chalcogens causes optical excitations of their electrons. The principal type of optical excitation that occurs is the collective oscillation of electrons in the valence band of the metal.

3.2 Summary

Novel materials are urgently required for future needs. Chalcogens are in the Group 16 elements including sulfur, selenium, tellurium, and polonium. Recent advances to the design and fabrication of metal chalcogenides have created a wide range of novel materials possessing unique optical functionality that has enabled the production of traditional optical components or novel devices with features that are useful in applications such as memory storage and laboratory chip, translating in form from solely bulk geometries to include planar thin films and fibers. In addition, chalcogens are being exploited for biological, sensing, and energy technologies. Chalcogens and metal chalcogenides can meet certain optical design requirements, while their properties are inherently "fixed" by their crystalline nature. Chalcogenide glasses, on the other hand, are not confined to one specific composition, or atomic orientation; therefore, these materials can be tailored via "compositional design" that allows tunability of glass properties (thermal, mechanical, or optical) to yield novel functionality. Comparatively, the use of chalcogenides, which have infinite optical, thermal, electrical, and physical flexibility, can offer the benefits that can be attained through the use of several modern technologies such as sensors.

References

[1] X. Zhang, N. Peng, T. Liu, R. Zheng, M. Xia, H. Yu, et al., Review on niobium-based chalcogenides for electrochemical energy storage devices: application and progress, Nano Energy 65 (2019) 104049. Available from: https://doi.org/10.1016/j.nanoen.2019.104049.

[2] K. Chen, C. Wang, Z. Peng, K. Qi, Z. Guo, Y. Zhang, et al., The chemistry of colloidal semiconductor nanocrystals: from metal-chalcogenides to emerging perovskite, Coord. Chem. Rev. 418 (2020) 213333. Available from: https://doi.org/10.1016/j.ccr.2020.213333.

[3] C. Lin, C. Rüssel, S. Dai, Chalcogenide glass-ceramics: functional design and crystallization mechanism, Prog. Mater. Sci. 93 (2018) 1–44. Available from: https://doi.org/10.1016/j.pmatsci.2017.11.001.

[4] P. Lucas, M.A. Solis, D. Le Coq, C. Juncker, M.R. Riley, J. Collier, et al., Infrared biosensors using hydrophobic chalcogenide fibers sensitized with live cells, Sens. Actuators B Chem. 119 (2006) 355–362. Available from: https://doi.org/10.1016/j.snb.2005.12.033.

[5] X. Chen, R. Liu, S. Qiao, J. Mao, X. Du, Synthesis of cadmium chalcogenides nanowires via laser-activated gold catalysts in solution, Mater. Chem. Phys. 212 (2018) 408–414. Available from: https://doi.org/10.1016/j.matchemphys.2018.03.050.

[6] X. Lu, Z. Lai, R. Zhang, H. Guo, J. Ren, L. Strizik, et al., Ultrabroadband mid-infrared emission from $Cr2^{+}$-doped infrared transparent chalcogenide glass ceramics embedded with thermally grown ZnS nanorods, J. Eur. Ceram. Soc. 39 (2019) 3373–3379. Available from: https://doi.org/10.1016/j.jeurceramsoc.2019.04.048.

[7] J. Pan, Y. Qian, Synthesis of cadmium chalcogenide nanotubes at room temperature, Mater. Lett. 85 (2012) 132–134. Available from: https://doi.org/10.1016/j.matlet.2012.07.012.

[8] J. Li, Q. Sun, Z. Wang, J. Xiang, B. Zhao, Y. Qu, et al., NbSe3 nanobelts wrapped by reduced graphene oxide for lithium ion battery with enhanced electrochemical performance, Appl. Surf. Sci. 412 (2017) 113–120. Available from: https://doi.org/10.1016/j.apsusc.2017.03.249.

[9] R.A. Aziz, S.K. Muzakir, I.I. Misnon, J. Ismail, R. Jose, Hierarchical Mo9Se11 nanoneedles on nanosheet with enhanced electrochemical properties as a battery-type electrode for asymmetric supercapacitors, J. Alloy. Compd. 673 (2016) 390–398. Available from: https://doi.org/10.1016/j.jallcom.2016.02.221.

[10] Y. Xiong, N.C. Lai, Y.C. Lu, D. Xu, Tuning thermal conductivity of bismuth selenide nanoribbons by reversible copper intercalation, Int. J. Heat. Mass. Transf. 159 (2020) 120077. Available from: https://doi.org/10.1016/j.ijheatmasstransfer.2020.120077.

[11] I.R. McFarlane, J.R. Lazzari-Dean, M.Y. El-Naggar, Field effect transistors based on semiconductive microbially synthesized chalcogenide nanofibers, Acta Biomater. 13 (2015) 364–373. Available from: https://doi.org/10.1016/j.actbio.2014.11.005.

[12] S.H. Kim, K.N. Hui, Y.J. Kim, T.S. Lim, D.Y. Yang, K.B. Kim, et al., Oxidation resistant effects of Ag2S in Sn-Ag-Al solder: a mechanism for higher electrical conductivity and less whisker growth, Corros. Sci. 105 (2016) 25–35. Available from: https://doi.org/10.1016/j.corsci.2015.12.021.

[13] T.V. Moreno, L.C. Malacarne, M.L. Baesso, W. Qu, E. Dy, Z. Xie, et al., Potentiometric sensors with chalcogenide glasses as sensitive membranes: a short review, J. Non. Cryst. Solids 495 (2018) 8–18. Available from: https://doi.org/10.1016/j.jnoncrysol.2018.04.057.

[14] M. Soylu, A. Dere, C. Ahmedova, G. Barim, A.G. Al-Sehemi, A.A. Al-Ghamdi, et al., Investigating the coumarin capability in chalcogenide 20Tl2Se–80Pr2Se3 system based photovoltaics, Spectrochim. Acta Part. A Mol. Biomol. Spectrosc. 202 (2018) 123–130. Available from: https://doi.org/10.1016/j.saa.2018.04.075.

[15] N. Jia, M. Zhang, B. Li, C. Li, Y. Liu, Y. Zhang, et al., Ternary chalcogenide LiInSe2: a promising high-performance anode material for lithium ion batteries, Electrochim. Acta. 320 (2019) 134562. Available from: https://doi.org/10.1016/j.electacta.2019.134562.
[16] D. Hu, X. Wang, H. Yang, D. Liu, Y. Wang, J. Guo, et al., Host-guest electrocatalyst with cage-confined cuprous sulfide nanoparticles in etched chalcogenide semiconductor zeolite for highly efficient oxygen reduction reaction, Electrochim. Acta. 282 (2018) 877–885. Available from: https://doi.org/10.1016/j.electacta.2018.06.106.
[17] W. Xu, H. Liu, D. Zhou, X. Chen, N. Ding, H. Song, et al., Localized surface plasmon resonances in self-doped copper chalcogenide binary nanocrystals and their emerging applications, Nano Today 33 (2020) 100892. Available from: https://doi.org/10.1016/j.nantod.2020.100892.
[18] I. Csarnovics, M. Veres, P. Nemec, S. Molnár, S. Kökényesi, Surface plasmon enhanced light-induced changes in Ge-Se amorphous chalcogenide – gold nanostructures, J. Non. Cryst. Solids (2020) 120491. Available from: https://doi.org/10.1016/j.jnoncrysol.2020.120491.
[19] V.G. Orlov, G.S. Sergeev, E.A. Kravchenko, Bismuth and antimony chalcogenides: peculiarities of electron density distribution, unusual magnetic properties and superconductivity, J. Magn. Magn. Mater. 475 (2019) 627–634. Available from: https://doi.org/10.1016/j.jmmm.2018.12.001.
[20] C.C. Chang, T.K. Chen, W.C. Lee, P.H. Lin, M.J. Wang, Y.C. Wen, et al., Superconductivity in Fe-chalcogenides, Phys. C. Supercond. Its Appl. 514 (2015) 423–434. Available from: https://doi.org/10.1016/j.physc.2015.02.011.
[21] D. Rana, Rohit, S. Hussain, S.K. Mehta, D. Jamwal, A. Katoch, Metal chalcogenide nanomaterials based supercapacitors, Reference Module in Earth Systems and Environmental Sciences, Elsevier, 2020. Available from: https://doi.org/10.1016/b978-0-12-819723-3.00024-x.

CHAPTER

4

Gas sensors

4.1 Introduction

Rapid population growth, industrial activities, and vehicles are the primary sources of emissions released into the atmosphere, which frequently worsen natural environmental conditions from the local level to the global level. Amongst these emissions, the most hazardous are nitrogen dioxides (NO_2), sulfur dioxide (SO_2), carbon monoxide (CO), hydrogen sulfide (H_2S), and methane (CH_4). Monitoring these emissions continually is important to avert environmental degradation. Numerous devices are available for monitoring pollutants and hazardous gases; however, they are time-consuming, costly, and seldom utilized in real-time environments. Recent advancements in emission control technology have resulted in more efficient monitoring and measuring systems, including sensor-based systems, that may be utilized to comply with rigorous emission regulations. A gas sensor is a system that senses the presence or concentration of gases in the atmosphere. By changing the resistance of the material within the sensor in response to the gas concentration, the sensor generates a corresponding potential difference, which can be calculated as output voltage. Due to its numerous uses in a variety of disciplines, including environmental monitoring, medicine, and industrial process monitoring, gas sensors have garnered considerable interest [1]. Several gas sensors have been described to date, including chromatographic sensors [2], electrochemical sensors [3], and spectrometric sensors [4]. Today, there is an increasing demand for gas sensors with focused selectivity for specific gases in a variety of fields of application, including industrial tasks such as process control, homeland security challenges such as the detection of chemical or biological hazardous substances, and the growing field of building automation and pollution control. All of these applications require the gas sensors

Metal Chalcogenide Biosensors
DOI: https://doi.org/10.1016/B978-0-323-85381-1.00007-6

to operate reliably, which means that in addition to selectivity, long-term stability and resistance to humidity are important qualities for future gas detecting components. Low power usage is also the desired attribute in light of the expected increase in energy costs.

Accurate gas sensing is a critical application in the field of emissions monitoring these days. Spectroscopic sensing techniques are the most widely used, but they need large, complicated, and expensive setups. Sensors for integrated in-situ and real-time monitoring have become an alternative that could also offer a downsizing and low-cost solution. Over the last few decades, numerous types of gas sensors have been developed utilizing a variety of materials and transduction platforms. These include carbon nanotubes, graphene, metal/metal oxide nanoparticles, two-dimensional nanomaterials, semiconductor, and hybrid nanostructures [5].

Amongst these, metal oxide gas sensors of the Taguchi type dominate the market for gas sensors [6]. While the sensor's response is proportional to the gas concentration, the sensor's sensitivity and selectivity are entirely dependent on the material used and the operating temperature. Despite their widespread usage, metal oxide gas sensors have several drawbacks, including limited selectivity for specific gases, response and recovery times measured in minutes or hours, and a significant effect of humidity on the sensor output. Concerning this, the prospect of employing chalcogenide glassy semiconductors as the sensitive layer in gas sensors for the study of industrial solutions and pollutant gases has received substantial interest in recent years. Significant progress may be made by examining the conductivity variations of these layers as a sensitive metric. Conductivity measurements are one of the oldest and most sensitive surface physics techniques. Adsorption activities, as well as changes in the surface structure, result in significant variations in resistance. A plethora of approaches, including chalcogenide-based gas sensors photoelectric can be employed to detect emissions for environmental monitoring applications [7] pioneered research on the electrical characteristics of chalcogenide-based materials of Te films when oxygen and nitrogen were diluted in either dry or wet air. The results indicated that the concentration of holes increased during the adsorption process but not their mobility. However, the changes in electrical characteristics produced by these gases were extremely tiny and irreversible, as later confirmed and explained in a comprehensive, pertinent investigation [8] began a detailed investigation of the usage of chalcogenide-based thin films as an active element in gas sensor production in the early 2000s.

Nowadays, in comparison to oxide or fluoride-based materials, chalcogenide-based materials are attracting a lot of attention from researchers in various fields because of their promising features including high refractive index, transparency in the mid and far infrared region, and low phonon energy [9]. In these fields, the qualities of these materials played an indispensable role in ensuring the long-term reliability of device performance and use. They may be doped with rare earth elements such as erbium (Er), neodymium (Nd), and praseodymium (Pr), making them suitable in a variety of environmental applications, both active and passive. Pure chalcogenide-based materials thin films such as Te have been demonstrated to be chemically sensitive to various gases and vapors [10]. This chapter discusses the efforts undertaken over the years concerning these materials' sensing activity for monitoring the concentrations of nitrogen oxides, carbon oxides, hydrogen sulfide, and methane. In this chapter, the fundamental background of emissions, sources, impact on the environment and health, and emissions control technologies are explained to give insight. Additionally, potential prospects for fabricating high-performance gas sensors employing these materials are addressed.

4.2 Emissions

Emission is one of the biggest scourges of our day, not only because it influences climate change but also because of its impact on public and individual health due to increased illness and death. Many contaminants are significant contributors to human illness. Emission is defined as the production and discharge of substances into the atmosphere. These emissions can be in the form of particles, gas, or radiation and are harmful to the environment. When released, they can modify the natural characteristics of the atmosphere and are referred to as "pollutants." These pollutants can be generated by both natural processes and human or anthropogenic activities. These pollutants are categorized into two types: primary and secondary. Primary pollutants are substances that are formed directly by a process, such as volcanic ash or carbon monoxide gas from automobile exhaust. Primary pollutants can be transformed into various chemical species through atmospheric reactions. These reactions can create both benign compounds and secondary air pollutants that are potentially more dangerous than their precursors. In another word, secondary pollutants are created when other pollutants (primary pollutants) react in the atmosphere, rather than being directly emitted. All these pollutants are releasing a wide range of species into the free atmosphere, putting a strain on the ecosystem's natural ability to adapt or compensate for the increased concentrations.

Some emission causes may be linked to both natural and human-caused events. These emissions can harm the atmosphere by interfering with climate, plant and animal physiology, whole habitats, and human property, such as crops and man-made structures. Table 4.1 lists the world's most significant air pollutants or emissions and their origins.

TABLE 4.1 Lists the world's most significant air pollutants or emissions and their origins.

Pollutant	Natural source	Anthropogenic source
Nitrogen oxides (NO_x)	Naturally occurring nitrogen oxides are generated by lightning and, to a lesser extent, by microbiological activities in soils	The primary sources of nitrogen oxides produced by humans are road transport and the public power and heat industry
Particulate matters (PM)	PM is derived naturally from sea salt, dust (such as airborne dirt), and pollen, but it also contains material from volcanic eruptions and particles produced from naturally occurring gaseous precursors (e.g., sulfates)	Agricultural operations, industrial processes, burning of wood and fossil fuels, building and demolition activities, and entrainment of road dust into the air are all examples of these sorts of activities
Sulfur dioxide (SO_2)	Volcanic eruptions and decay can be a substantial source of sulfur dioxide emissions in nature	Sulfur dioxide gas is mostly produced by humans via the burning of fossil fuels in power plants and industrial operations
Carbon monoxide (CO)	Unnoticeable	Rich and stoichiometric combustion, mainly from vehicles
Carbon dioxide (CO_2)	Ocean outgassing, decaying plants and other biomass erupting volcanoes, naturally occurring wildfires, and even ruminant animal belches	Power generation, transportation, industrial processes, chemical, and petroleum manufacturing, and agricultural operations
Non-methane hydrocarbons	Biological processes	Fuel production, distribution, and combustion, with the main sources being emissions from motor vehicles as a result of gasoline evaporation or incomplete combustion, and biomass burning
Methane (CH_4)	The greatest source is natural wetlands, which release CH_4 because of microorganisms decomposing organic matter in the absence of oxygen. Termites, seas, sediments, volcanoes, and wildfires are all smaller sources	Landfills, oil and natural gas pipelines, agricultural operations, coal mining, stationary and mobile combustion, wastewater treatment, and some industrial processes are all examples of industrial processes

(*Continued*)

TABLE 4.1 (Continued)

Pollutant	Natural source	Anthropogenic source
Chlorofluorocarbons (CFC)	None	Numerous applications include air conditioning, refrigeration, blowing agents in foams, insulations, packaging materials, aerosol propellants, and solvents

Volcanic eruptions and wind erosion are examples of emissions from natural sources. Volcanic eruptions can introduce very important quantities of gases and particles into the atmosphere. For example, according to a report by [11], the Etna volcano emits 3000 tons of SO_2 on an average day and up to 10,000 tons during periods of great activity. During the cataclysmic eruptions of the Tambora in 1815 in Indonesia, 100 billion tons of volcanic products were ejected into the atmosphere, 300 million tons of which reached the stratosphere, which resulted in a mean temperature fall of $\sim$0.7°C over the whole Earth. In some rural areas, periodic forest fires produce large amounts of particulate matter. Other natural sources of atmospheric emissions include thunderbolts, which produce significant quantities of NO_x; algae on the surface of the oceans, which give out H_2S; wind erosion, which introduces particles into the atmosphere; and humid zones, such as swamps or little deep lakes, which produce CH_4. Low concentrations of ozone (O_3) occur naturally at ground level, formed in the presence of sunlight by reactions between NO_x and volatile organic compounds (VOCs).

Anthropogenic activities or human-caused events trigger a large number of emissions in the atmosphere, which is one of the world's most important public health hazards, resulting in around nine million deaths each year [12]. Internal combustion engine emissions are a prime example of anthropogenic emissions that can be classified as mobile or point sources. Examples of mobile sources include vehicle emissions such as automobiles, trucks, airplanes, and marine engines, while point sources include factories, industrial areas, and power generation plants. Many researchers believe, that the dawn of the industrial age and a big success in terms of technology, society, and the provision of a wide range of services, also resulted in the release of massive amounts of pollutants into the air that is hazardous to human health.

Vehicle emissions are currently the most significant cause of pollutants in the atmosphere, especially in urban areas. According to [13]. Vehicle emissions have been a source of concern since the early 1950s when a researcher discovered that traffic was to blame for the smoggy skies over Los Angeles USA. At that time, typical new cars were emitting nearly 13 g of hydrocarbons per mile, 3.6 g of nitrogen oxides per mile, and 87 g of CO per mile. Carbon-based fuels (coal, fuel oil, wood, natural gas) are never completely burned, resulting in the production of carbon monoxide (CO) and hydrocarbons. When fossil fuels in motor fuel are burned at high temperatures, a mixture of nitrogen and oxygen is produced. Human activities have also resulted in an increase in VOCs from the petroleum, chemical, and transportation industries, as well as NO_x from combustion and power generation plants. As a result, O_3 is more concentrated, and smog occurs more frequently in highly inhabited and industrial areas. Human activities add to the overall amount of ambient particulate matter. Particles are generated mostly in urban areas as a result of combustion from these sources. CO_2 is formed when coal and sulfur from fuel oils oxidize. These are the fuels that are utilized to transport, warm, and supply energy to several industrial operations. These emissions have been demonstrated to have a range of harmful health and environmental consequences. Table 4.2 lists the main types of emissions emitted by combustion and power generation plants, along with an explanation of their impact.

It is important to note that all man-made sources contribute to the deterioration of air quality in addition to natural sources that have existed since the Earth's creation and are in direct dependency on it. Humans began dramatically increasing atmospheric emissions, enhancing the planet's natural greenhouse effect, and resulting in increased temperatures, which resulted in climate change. From this, we can see that anthropogenic activities contribute to the greenhouse effect, which results in global warming. Global warming wreaks havoc on ecological systems. As permafrost ice melts, the seal level tends to rise. Changes in global precipitation patterns have been seen with an increase in dry areas and a decrease in regional floods, as well as an increase in the frequency of severe occurrences such as tsunamis, earthquakes, cyclones, and tornadoes. This major issue is also intertwined with social, economic, and political considerations, as well as lifestyle choices. Urbanization and industrialization are reaching unprecedented and unsettling levels over the world.

4.2.1 Overview of emission control technologies

With the recent focus on global climate change caused by greenhouse gases, one could worry about the future of a kind of equipment that produces CO_x and NO_x as a desirable end products. The aims have shifted away from merely eradicating target emissions and toward minimizing these hazardous combustion byproducts. The amount and

TABLE 4.2 Emissions released by combustion and power generation plants and their impacts.

Emission	Description	Impacts on health and environment
Nitrogen oxides (NO_x)	Produced when nitrogen in the air combines with oxygen at the engine's high temperature and pressure in all types of combustion facilities, from large to small including motor engines, and furnaces. NO_2 is a highly reactive gas. Because NOx production increases when an engine is operated at its most efficient operating point, there is a natural trade-off between efficiency and NOx emission control. It is projected to be significantly decreased with the usage of emulsion fuels	It has the potential to cause respiratory difficulties in both people and animals. In high-altitude atmospheres, it can form acids
Carbon monoxide (CO)	A byproduct of incomplete combustion	Upon inhalation, poses a significant risk to the health and lives of animals and humans. It can impair the blood's ability to carry oxygen; excessive exposure (carbon monoxide poisoning) can be lethal
Particulate matter (PM)	Particles in the air (fly ash, sea salt, dust, metals, liquid droplets, and soot) come from a variety of natural and human-made sources. Particulates are emitted by factories, power plants, vehicles, etc., and are formed in the atmosphere by condensation or chemical transformation of emitted gases	PM is responsible for a variety of environmental impacts, including acid precipitation, and damage to plant life. Particulate matter has several adverse health impacts, including respiratory disorders and cancer. Cardiovascular disease has been linked to extremely small particulate particles
Sulfur dioxide (SO_2)	It is a highly corrosive gas that is damaging to human health and the environment. It is produced through the combustion of coal and fuel oil, the smelting of nonferrous metal ores, the refining of oil, and the creation of electricity	It irritates the respiratory system, harms green plants, and is a precursor to acid precipitation

(*Continued*)

TABLE 4.2 (Continued)

Emission	Description	Impacts on health and environment
Volatile organic compounds. (VOCs)	Organic compounds with a boiling point of less than or equal to 250°C. produced during the burning of hydrocarbons in engines	Have a negative influence on the atmosphere; they obstruct the production of stratospheric ozone
Chlorofluorocarbons (CFC, HCFC, HFC)	Typically, these are refrigerants or propellants	They may contribute to the stratospheric ozone layer's depletion and the greenhouse effect
Hydrocarbons (HC)	Hydrocarbons are a toxic class of fuels that have been burned or are in the process of being burned	Hydrocarbons contribute significantly to smog, which can be a significant concern in urban areas. Prolonged contact with hydrocarbons contributes to the development of asthma, liver disease, lung disease, and cancer

content of process emissions from batch or continuous processes might vary significantly. Flares, vapor combustors, thermal oxidizers, and catalytic systems have all been used to destroy emissions at a specific facility, depending on the process conditions, destruction needs, and energy demands. In addition, clean air requirements that are becoming increasingly demanding are pushing advancements in manufacturing technologies. These manufacturers are adapting to the shifting air emissions management scenario, and those that do so will reap environmental and economic rewards.

In the emission control technologies of NO_x, pre-combustion is controlled by reducing the nitrogen concentration of the fuel by refining. NO_x emissions are regulated in combustion modification by changing the operating parameters to obtain the lowest feasible concentration. There are several techniques to reduce NO_x such as Selective catalytic reduction (SCR), Selective non-catalytic reduction (SNCR), and Plasma assisted NO_x reduction technology. SCR is a well-established technique for NO_x abatement due to the variety of processes and catalysts [14]. With moderate temperature ranges from 200°C–500°C, the technique requires higher initial investment and ongoing operating costs [15]. This method is capable of removing more than 75% of NOx [16]. The SCR system catalytically transforms NO_x in the flue gas stream to N_2 and H_2O. The SCR process utilizes a variety of reducing agents, including ammonia/urea, hydrocarbons, and hydrogen. The SNCR method, also known as Thermal $DeNO_x$, is used to decrease NO_x without using a catalyst at elevated temperatures (900°C–1000°C). Ammonia and urea are the most often utilized reducing agents in this procedure. The chemical processes are identical to those of SCR, with the exception that they include more complicated chemistry and free radical reactions. The total reaction for NO_x reduction in the case of urea as a reducing agent is.

Plasma can contribute significantly to the reduction of NO_x emissions from internal combustion engines due to its active species, which exhibit a wide variety of unique properties and can catalyze or enhance the chemical reaction involved in the process. Additionally, plasma can burn materials that are above their flammability limits due to the fast production of thermally activated species. Due to the above-mentioned characteristics and the critical role of NO_2 conversion in the selective catalytic reduction (SCR) process, plasma has been extensively utilized to decrease NO_x via NO_x oxidation to NO_2.

In the post-combustion CO_2 capture and reduction variety of techniques can be used, including adsorption. Adsorption, using adsorbents such as metal-organic frameworks (MOFs) and zeolites are cost-effective, ecologically friendly, resistant to moisture and chemicals, and have a large surface area. This technique can be used to reduce CO_2. On the other hand, SO_x emissions can be reduced by using low sulfur fuels or by utilizing a variety of methods, including flue gas desulfurization (FGD), wet flue gas desulfurization (WFGD), dry sorbent injection (DSI), and bio-desulfurization.

In controlling the emissions from vehicle engines, catalytic converters and particulate filters are frequently employed in conjunction with an integrated approach to emissions management that includes the combustion system, the quality of the fuel and reductants, and electronic control systems are applied. They are promising approaches for achieving up to a 90% reduction in engine exhaust pollutants [17]. A catalytic converter is a type of exhaust emission control device that catalyzes a redox process to convert harmful gases and emissions in exhaust gas from an internal combustion engine to less dangerous pollutants. Catalytic converters have been used as emission control technology since the 1950s. They are often used in internal combustion engines that run on gasoline or diesel, particularly lean-burn engines, and occasionally in kerosene heaters and stoves. There are various types of catalytic converters such as three-way catalytic converter, diesel oxidation catalyst, selective catalytic reduction system, lean NO_x catalyst, lean NO_x trap, and non-selective catalytic reduction system while catalytic converters are most frequently found in vehicle exhaust systems, they are also used in electrical generators, forklifts, mining equipment, trucks, buses, locomotives, motorbikes, and ships. They are even used to reduce pollutants on some wood burners. Since the early 1980s, three-way catalytic converter catalytic converters have been the dominant pollution control technology for light-duty gasoline cars. When combined with an oxygen sensor-based, closed-loop fuel supply system, this technique enables simultaneous conversion of the three criterion pollutants generated during the combustion of fuel in a spark-ignited engine, HC, CO, and NO_x. On the interior walls of the honeycomb substrate, the active catalytic materials are coated with a thin layer of precious metals such as Pt, Pd, and Rh, as well as oxide-based inorganic promoters and support materials. Typically, the substrate has a high number of parallel flow channels to provide sufficient contact area between the exhaust gas and the active catalytic materials without introducing excessive pressure losses.

The particulate filter can physically capture diesel particles and prevents them from being discharged from the exhaust. In the mid-1970s, diesel particulate filters were introduced for use in diesel vehicles. Diesel particle filters are used to avoid hazardous pollutants such as NO_x from the diesel engine's combustion. Diesel particulate filters, which are typically constructed of a specific ceramic material with a temperature resistance of 1200°C, are also available in silicon carbide and cordierite variants [18].

The regenerative thermal oxidizer is a commonly utilized abatement device in industrial air emissions management applications because it can reuse up to 97% of the thermal energy generated during combustion to pre-heat entering, untreated pollutants. In operation, the solvent-laden air enters one of the regenerative thermal oxidizer energy recovery chambers, where it is preheated by the high-temperature ceramic heat transfer media before being introduced into the oxidation chamber. The temperature of the solvent-laden air quickly increases as it travels through the bed. Following the purification procedure by chemical oxidation, the hot, clean departing gas warms the outlet energy recovery bed. To ensure the beds' maximum heat recovery effectiveness, the automated diverter valves switch the solvent-laden airflow direction at regular intervals under command from the programmable logic control system. This periodic shift results in a temperature distribution that is consistent throughout the oxidizer. With a sufficient concentration of hydrocarbons in the process air stream, the heat energy contained in the emissions will self-sustain the oxidation process, eliminating the need for additional heat energy.

4.3 Chalcogenide-based gas sensors for various gas emissions

Chalcogenides are a significant class of materials that include one or more of the chalcogen elements with unique characteristics, namely sulfur (S), selenium (Se), and tellurium (Te). The sixteen group elements of the periodic table are most frequently fused with other elements such as As, Ge, P, Sb, Bi, Si, Sn, Pb, Al, Ga, Ag, and La to form chalcogenide glass materials with a variety of remarkable characteristics. The most fascinating family of inorganic materials is tungsten and molybdenum chalcogenides, which are composed of metal oxides (MO_3; M = Mo and W) and metal dichalcogenides (MX_2; X = S, Se, and Te). They have been intensively explored in several applications due to their intriguing surface characteristics and simplicity of manufacturing, including lithium and sodium ion storage, supercapacitors, gas sensors, and biosensors. A comprehensive review of chalcogenide-based glass materials and their applications is discussed in Ref. [19]. Recent advances in chemiresistive gas sensing applications with chalcogenide-based materials are presented in Ref. [20].

Since the discovery of the impact of propylamine adsorption on dimorphite (As_4S_3) conductance, the potential for detecting hazardous gases using chalcogenide semiconductors have been actively studied. Dimorphite is a very rare orange-yellow arsenic sulfide mineral with the chemical name arsenic sesquisulfide (As_4S_3) which is originally thought to be one of two dimorphous substances. Dimorphite is formed mainly in volcanic fumaroles at temperatures of 70°C–80°C (158–176°F) in nature. In the mid-19th century, dimorphite, one of the four recognized chemical compounds of the As–S system, was discovered as a mineral [21]. Dimorphite crystallizes in two distinct forms, A- and B-. Since dimorphite is a molecular crystal, it can be synthesized by melting the required components in a vacuum. Structural studies by [22] revealed the presence of two phases of the same Pnma space group, (dimorphite I) and (dimorphite II). Dimorphite can be formulated in a vacuum by melting arsenic and sulfur in the appropriate molar ratios. Due to dimorphite's semiconductive properties, preliminary research indicates that it could be used in the manufacture of gas sensors [23–25]. Marian et al., [23] presented a study on dimorphite-based gas-sensitive thin films. The chemical sensor for the detection of propylamine vapor was a sandwich metal-semiconductor-metal structure with dimorphite as the semiconducting material. The melt-quenching method was used to make polycrystalline dimorphite ingots from pure As and S in evacuated quartz ampoules. Sandwich samples were created by depositing aluminum, dimorphite, and aluminum on glassy substrates in a series of vacuum (103 Pa) depositions. The substrate temperature was held at 333K and dimorphite film growth velocities were 0.2 or 5.0 m/h. The films had a thickness of 7–8 m. The effects of gas on current-voltage characteristics and transient characteristics were investigated. In their research, they discovered that thin films made of artificial dimorphite (As_4S_3) were gas sensitive at room temperature. The interaction of propylamine vapor with dimorphite caused a rise in current, indicating a doping effect caused by the gas. The gas-induced current was found to be dependent on both the applied voltage and the gas concentration. Thin films made of dimorphite were sensitive to propylamine vapor at various bias voltages and concentrations.

This section discusses the efforts undertaken over the years concerning these materials' sensing activity for monitoring the concentrations of NO_x, CO_x, SO_x, and CH_4.

4.3.1 Nitrogen oxides

4.3.1.1 *Overview of* NO_x

Nitrogen oxides (NO_x) represent a family of seven compounds that are highly toxic gases formed during the combustion of nitrogen and oxygen at high pressures and temperatures. NO_x is constituted primarily of nitric oxide (NO) and a trace amount of more toxic nitrogen dioxide (NO_2) (Fig. 4.1). Table 4.3 summarizes the family of NO_x compounds and their characteristics.

The most prevalent gases in the atmosphere are N_2O, NO, and NO_2. N_2O is abundantly generated by biological sources such as plants and

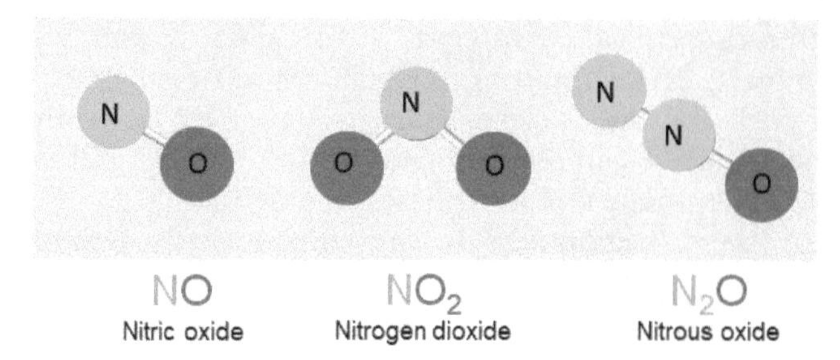

FIGURE 4.1 Nitrogen oxides (NO_x).

TABLE 4.3 Family of NO_x compounds and their characteristics.

Formula	Name	Properties
N_2O	Nitrous oxide	Colorless gas water soluble
NO N_2O_2	Nitric oxide dinitrogen dioxide	Colorless gas is slightly water soluble
N_2O_3	Dinitrogen trioxide	Black solid water-soluble decomposes in water
NO_2 N_2O_4	Nitrogen dioxide dinitrogen tetroxide	Red-brown gas is very water soluble and decomposes in water
N_2O_5	Dinitrogen pentoxide	A white solid is very water soluble and decomposes in water

yeasts. It is a slightly irritant and analgesic. NO_2 is a significant air pollutant by itself and also can interact with ozone (O_3) and acid rain in the environment. At any temperature, oxidation of NO_2 by O_3 produces both molecular oxygen (O_2) and either NO or two NO molecules linked together as dinitrogen dioxide (N_2O_2). NO or N_2O_2 is then rapidly oxidized in about two hours, to NO_2. When a photon of ionizing radiation from sunlight strikes NO_2, it converts a molecule of oxygen (O_2) into an ozone molecule [26]. N_2O is also a greenhouse gas, similar to carbon dioxide (CO_2), in that it absorbs long-wavelength infrared light to trap heat emitted by the Earth, contributing to global warming. N_2O has an estimated atmospheric lifespan of around 150 years. Its major mode of destruction is by transit to the stratosphere, where it absorbs ultraviolet light, forming nitrogen dioxide and an oxygen atom. Dinitrogen trioxide (N_2O_3) and dinitrogen tetroxide (N_2O_4) occur in trace amounts. They do, however, exist in such low quantities in the atmosphere that their presence and influence are frequently overlooked. N_2O_4 is a dimer of two NO_2 molecules and reacts similarly to NO_2; hence, the existence of N_2O_4 may be obscured by the more prevalent NO_2 [27].

The majority of NO_x emissions from combustion are in the form of NO. NO is produced more than the available oxygen in the air, about 200,000 ppm at temperatures more than 1300°C [28]. Except for NO produced by soils, lightning, and natural fires, the majority of NO is anthropogenic [29]. Natural or biogenic sources are usually believed to be responsible for fewer than 10% of total NO emissions. NO causes the same inability of the blood to take oxygen as carbon monoxide (CO). Since NO is very minimally soluble in water, it offers no significant hazard to humans save for newborns and extremely sensitive persons.

Vehicles and road traffic sources account for roughly half of the NO_x emissions. Boilers in electric power plants account for approximately 40% of NO_x emissions from stationary sources [28]. Additionally, anthropogenic sources such as industrial boilers, incinerators, gas turbines, reciprocating spark ignition and diesel engines in stationary sources, iron and steel mills, cement and glass manufacturing, petroleum refineries, and nitric acid manufacturing contribute significant emissions. Lightning, forest fires, grass fires, trees, shrubs, grasses, and yeasts are all examples of biogenic or natural sources of nitrogen oxides. These diverse sources generate varying amounts of each oxide.

It can be seen that NO_x is one of the most hazardous gases produced during combustion, industrial operation, and vehicle operation. The detection and control of these emissions are critical; thus, much effort has been made to create sensors for monitoring the NO_x content in the ambient atmosphere in a variety of settings including those based on semiconductors, solid electrolytes, and organic materials [5]. Most of these studies aim at determining critical parameters like sensitivity, selectivity, response–recovery periods, and required operating temperature. It is not easy to design a sensor with acceptable parameters; hence, significant effort is being made to optimize device performance by empirical methods such as doping, fabrication of nanoscale thin films, annealing, sol-gel, and liquid phase deposition technologies [30].

4.3.1.2 *Chalcogenide-based gas sensor studies for NO_x*

Throughout the literature, we can see that the recent technical developments have had negative consequences for humanity, including NO_x pollution because of increasing industrialization and urbanization. Subsequently, effective sensors are necessary to initially differentiate these poisons. A lot of efforts are being made to develop efficient chalcogenide-based gas sensors for monitoring the NO_x content.

In a study by [24], the glassy alloys were made by melting pure (99.99%) As, S, Te, and Ge in quarts of ampoules. They presented the results of gas sensing measurements carried out at room temperature using impedance spectra in the 5–13 MHz frequency range. To test the use of these materials in future gas sensors operating at room temperature (22°C), the impedance spectra of quaternary As–Ge–S–Te dependent alloys were studied in both dry synthetic air and a mixture of nitrogen dioxide. $As_2Te_{13}Ge_8S_3$ and $As_2Te_{130}Ge_8S_3$, as well as pure Te, to obtain data on the effect of tellurium were used. In the study, the glassy alloys were made by melting pure (99.99%) As, S, Te, and Ge in quarts of ampoules at a pressure of 5 105 Torr. At 900°C, the melting took place. During the synthesis time (24 h), the ampoule was rotated around the longitudinal axis at a speed of 7–8 rotations per minute and agitated for homogenization before being quenched on a copper refrigerator with running water. Two quaternary compositions, $As_2Te_{13}Ge_8S_3$ and $As_2Te_{130}Ge_8S_3$, were synthesized, as well as pure polycrystalline tellurium. Thermal "flash" evaporation of priory-synthesized materials from tantalum boats onto ceramic Al_2O_3 substrates containing previously deposited platinum interdigital electrodes with an electrode diameter of 15 m and interelectrode lengths of 45 m created the chalcogenide thin films (Fig. 4.2). At room temperature, quaternary chalcogenides in the device As–Ge–S–Te showed a high impedance sensitivity to NO_2 gas molecules in the air. These findings demonstrated how the film's structure and phase-structural state influence not only the mechanism of current flow, but also their gas sensing properties and, ultimately, their ability to be used in the production of room-temperature gas sensing instruments.

A ridge Ge–Sb–Se chalcogenide glass waveguide sensor with a power confinement factor (PCF) of 5% was suggested to detect NO_2 [31]. This sensor enabled sensitive detections as well as potential N_2O detection limits of 1.6 ppm at 7.7 m, which were lower than the

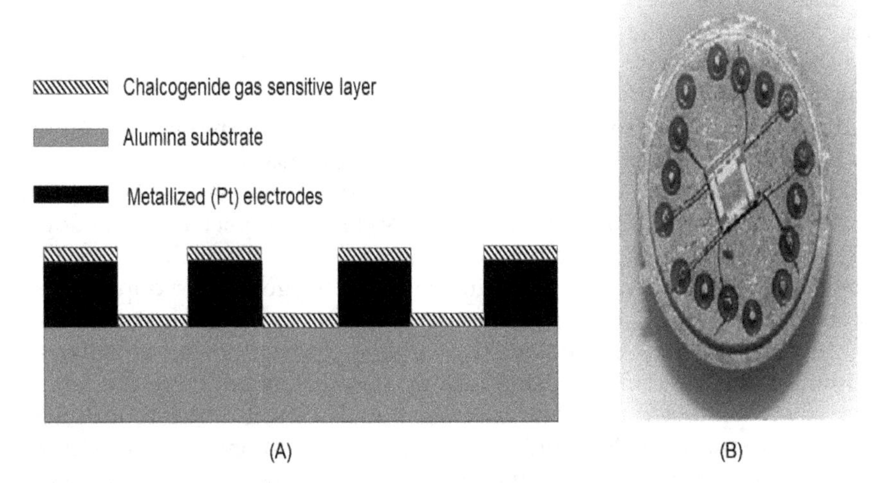

FIGURE 4.2 Schematic representation of (A) chalcogenide-based thin film device and (B) its view encapsulated in standard socket [24].

international environmental standards' occupational exposure limits. As a result, our theoretical and practical findings represent the initial step in the creation of a mid-infrared chalcogenide optical integrated sensor.

In an earlier study by [23] the characterization of chalcogenide-based thin films as gas-sensing materials was presented. The sensing behavior of the As–S–Te films was tested with environmental pollutant gases such as NO_2, CO, and SO. Nitrogen dioxide has a high degree of sensitivity. NO_2 had a detection range of 0.95–1.9 ppm in ambient air. The reaction and recovery times are short, with a high degree of repeatability and sensitivity. At room temperature, all measurements were made. Applications using gas sensing are discussed. Gas sensors based on As–S–Te ternary alloys demonstrate a high sensitivity to NO_2 at low concentrations. This sensitivity is not only dependent on the gas concentration but also on the bias voltage applied. In addition to this, it has been shown that these sensors exhibit a small sensitivity to other pollutant gases in the air, such as SO_2 and CO These sensors also worked at room temperature and had high sensitivity in the ppm and sub-ppm levels, as well as a high degree of selectivity. These findings demonstrated that chalcogenide-based gas sensors represented a novel class of solid-state, low-cost, and low-power chemical sensors that are particularly well-suited for environmental monitoring.

The binary, ternary, and quaternary alloys of Te, Ge, Se, As, Mg, I, and S based on $Te_{13}Ge_x$ were studied in [10] order to determine acceptable material compositions for future room temperature gas sensors. While the majority of the alloys tested demonstrated a linear relationship between the NO concentration and the sensor current at constant voltage, the composition of the alloys had a significant effect on the NO sensitivity and signal/noise ratio. For instance, when As and Mg are combined, the NO sensing characteristics of $Te_{13}Ge_x$ films are reduced in comparison to films containing S or I. Among the alloys examined, two potential candidates for future gas sensors have been identified: $Te_{13}Ge_3S_3$ and $Te_{13}Ge_3$. The observed sensitivity was mostly due to the incorporation of Te but was also highly impacted by the doping and/or stabilizing components utilized. While the inclusion of Ge stabilized the sensitive layers against ambient oxygen, other doping elements such as As, S, or Mg reduce sensitivity. Due to the instability of the doped material, the potential of iodine doping can be ruled out. Although the basic mechanisms behind these behaviors were not studied, it was predicted that their investigation disclosed routes to individually customized chalcogenide gas sensors. Additionally, for usage in commercial gas sensors, additional efforts must be made to enhance long-term stability and repeatability.

One of the common chalcogenide compounds and alloys is tellurium (Te) which poses unique physical and multifunctional chemical characteristics [32] that make it ide [8,33–36]. Te also has been a popular element with a wide range of biological applications over the last decade due to its potential to be utilized as quantum dots in imaging and diagnostics and its antibacterial characteristics [37]. The inclusion of germanium may be used to stabilize the gas sensing characteristics of Te-based sensor layers. The metallic character of pure Te is diminished in TeGe alloys due to electron localization inside a glassy network that is stable for Ge concentrations below 20–25 at.%. Increased Ge concentrations result in crystallization and give the material a more metallic appearance [38]. Due to the decreased metallic nature of these compositions, the electric layer resistance increases. It is expected that the fundamental mechanism of interaction mentioned before for pure Te layers also happens in Te alloys.

Te has been identified as [8,39,40] described the creation of ultrathin tellurium-based sensors capable of detecting NO_2 gas in the ppm range at room temperature. They discussed the development of thin film tellurium-based sensors capable of detecting NO_2 gas in the ppm range at ambient temperature. These characteristics make them well-suited for environmental monitoring applications. Gas sensors based on Te thin films, in particular, are easily produced through thermal evaporation of pure tellurium and exhibit no stoichiometric flaws. However, it has been discovered that the gas sensing capabilities of these films are highly dependent on deposition factors such as substrate temperature, annealing conditions, film thickness, and substrate type.

In another study by [8], the effect of temperature and annealing on the electrical and sensing characteristics of tellurium-based films toward NO_2. The results were addressed in terms of the contributions to total conductivity made by the grain border, grain mass, and surface resistance. It was thought that the hole-enriched area on the surface, including the grain boundary, formed as a result of the dangling bond chalcogen's lone-pair electron interaction. Due to the interaction of NO_2 molecules with lone-pair electrons, the chemisorption of these molecules resulted in hole enrichment of the surface and grain boundary area.

For NO_2 gas sensor applications, Te thin films were formed onto glass and alumina substrates using an rf sputtering technique [41]. Scanning electron microscope (SEM), X-ray diffraction, and Raman spectroscopy were used to study the surface morphology and structure of the deposited films. As the film thickness grew from 100 to 300 nm, the morphology of the films changed from smooth to polycrystalline. The NO_2 gas sensing capabilities of the films were established by measuring their resistance change as a function of operating temperature and gas concentration. The surface shape of the films had a significant effect on the gas sensitivity. At room temperature, films with a thickness of 300 nm produced on a glass substrate exhibited the highest sensitivity to NO_2. Additionally, they evaluated the response and recovery times, repeatability, and long-term stability of these films. A sensor structure consisting of a TeO_2/SnO_2 p–n heterointerface was developed using the rf sputtering approach and effectively used for the detection of NO_2 gas at trace levels (1–10 ppm) at a significantly lower working temperature of 90°C [42]. Integration of an n-type SnO_2 sensing layer with p-type TeO_2 microdisks (600 m diameter) with an optimal thickness of 18 nm increases by sensing response for 10 ppm NO_2 gas at a relatively low working temperature (90°C) and a rapid reaction time (1.25 min). The heterointerface sensor structure's superior sensing response characteristics are attributable to the development of p–n junctions at the interface of SnO_2 and TeO_2 microdisks, as well as variation of the depletion width upon interaction with the target NO_2 gas. The origins of the heterointerface sensor's increased sensing mechanism have been examined in length. On the other hand, [34] examined the gas sensitivity of thermally evaporated Te films using a variety of substrates, deposition temperatures, and post-deposition annealing. The films were deposited on a variety of substrates at temperatures ranging from 77K to 373K. According to these investigations, the film formed at 373K on a glass substrate exhibited the highest sensitivity to NO_2.

One-dimensional nanostructure-based chemiresistive sensors have garnered considerable interest due to their compact design and superior sensing performance, which includes high sensitivity, low detection limits, low power consumption, and the potential to combine multisensor arrays. However, when operated at ambient temperatures, these sensors often exhibit poor response and recovery periods due to the slow catalytic or absorption/desorption processes. Therefore, attention has been made to the development of a chemiresistive sensor constructed with the chalcogenide-based Te nanotubes to overcome this obstacle which can work at a high temperature, which adds complexity to the device and raises its power consumption rate. Concerning this, Guan and his coworkers [43] developed an ultrasensitive room-temperature detection of NO_2 with a tellurium nanotube-based chemiresistive sensor to conduct NO_2 gas sensing, a customized sensing test instrument was made, which consists of a heating plate, rubber gasket, and a quartz cell. In their study, the detection of NO_2 measurements was made at room temperature. A lab view software was used to examine the sensor's real-time reaction. Different concentrations of the analyte gas were obtained by mass flow controllers mixing stoichiometric quantities of target gas and nitrogen. Typically, the gas sensing experiment was done at a pressure of 1 atm. The sensitivity of the sensor (S) was defined as:

$$S = \left[\left(R_g - R_o \right) / R_o \right] \times 100\% \tag{4.1}$$

where R_o and R_g are the sensor's electrical resistance in nitrogen and target gas, respectively. The recovery and response times were defined as the time necessary for the sensor to achieve steady-state and the time required to regain 90% of its maximum response, respectively. From the study, it was found that the constructed gas sensor displayed good sensitivity and selectivity for trace amounts of NO_2, with a detection limit of around 500 ppt. With the aid of UV irradiation, the gas sensor's reaction was completely reversible, and increasing the UV exposure decreased the sensor's recovery time to less than 5min. The excellent performance of this Te-based gas sensor was due to the nanotubes' high surface-to-volume ratio and crystallinity.

Zhang et al., [44] developed an ultrafast NO_2 sensor at ambient settings by using feather-like Te nanostructures functionalized on single-walled carbon nanotube networks. By fine-tuning the form and density of Te nanostructures, hybrid nanostructures exhibited good response and recovery durations of around 63 s and 7 min, respectively, to 100 ppbV NO_2 gas at ambient temperature. On the other hand, by combining electrospinning and galvanic displacement reaction, ultralong hollow tellurium (Te) nanofibers with regulated dimensions, shape, and crystallinity were synthesized [45]. Among the many nanofiber topologies, the branching Te nanostructure exhibits the highest sensing capacity for NO_2 at ambient temperature.

Vapor–solid growth of p-Te/n-SnO_2 hierarchical heterostructures and their enhanced room-temperature gas sensing properties by employing a two-step thermal vapor transport method was studied by [46]. In the study, they created brushlike p-Te/n-SnO_2 hierarchical heterostructures. When exposed to CO and NO_2 gases at ambient temperature, Te/SnO_2 hierarchical heterostructures altered resistance in the same manner as Te nanotubes and displayed significantly greater responses and quicker response rates than Te nanotubes. The improved gas sensing ability can be attributed to the increased specific surface areas and formations of many Te/Te or TeO_2/TeO_2 bridging point contacts, as well as the inclusion of p-Te/n-SnO_2 heterojunctions.

Quite recently, the effect of deposition rate and substrate microstructure on NO_2 gas sensitivity of Te thin films was investigated in Ref. [47]. Te thin films were deposited in a vacuum at various speeds (0.1–30 nm/s) on glassy, sintered alumina, and electrochemically nanostructured Al_2O_3 substrates. At room temperature, the sensitivity of manufactured films to NO_2 was determined. It is demonstrated that the deposition rate has a significant effect on the microstructure and gas sensing characteristics of the films under consideration. Increased deposition rate leads to the film's microcrystalline structure becoming amorphous. Both the gas sensitivity and reaction time decrease concurrently. Nanocrystalline and amorphous nanostructured tellurium (Te) thin films were then recently produced, and their gas sensing properties were studied [48] using scanning electron microscopy and X-ray diffraction analysis at various operating temperatures. Both kinds of coatings were found to interact with NO_2 resulting in a reduction in electrical conductivity. The gas sensitivity, responsiveness, and recovery periods of these two nanostructured films were different. Notably, these characteristics are also dependent on the operating temperature and gas concentration supplied to the films. Increases in operating temperature decreased not only the response and recovery times of the nanocrystalline films but also their gas sensitivity. This disadvantage may be overcome by employing amorphous nanostructured Te films, which, even at 22°C, displayed significantly greater gas sensitivity and faster reaction and recovery periods than nanocrystalline Te films. These findings were interpreted in terms of a rise in disorder (amorphization), which increased chalcogenide surface chemical activity, as well as an increase in the active surface area owing to substrate porosity.

As we know, the sensitivity and selectivity of a sensor are critical properties that are completely reliant on the material employed. The following Table 4.4 compares the NO_2 sensing properties of several Te materials [43]. As shown in the table, a chemiresistive sensor built using as-prepared Te nanotubes displays high sensitivity and selectivity for trace amounts of NO_2.

At ambient temperature, glassy tellurium chalcogenide compounds exhibit NO sensitivity. The observed sensitivity is mostly due to the

TABLE 4.4 NO_2 sensing properties of several Te materials.

Types of materials	Limit of detection	Sensitivity (% ppm^{-1})	References
Te thin film	0.75	−7.5 (1.5)	[30]
Te thin film	1	−1.8 (10	[41]
TeO_2/SnO_2 heterostructure	1	2200 (10)	[42]
TeO_2 thin film	50	−0.14 (100)	[49]
Te microtubes	10	−0.43 (200)	[49]
Te nanotube based chemiresistive sensor	5×10^{-4}	−5000 (1 ppb)	[43]

incorporation of tellurium but is also highly impacted by the doping and/or stabilizing components utilized. While the inclusion of germanium stabilizes the sensitive layers against ambient oxygen, other doping elements such as arsenic, sulfur, or magnesium reduce sensitivity. Due to the instability of the doped material, the potential of iodine doping can be ruled out. Although the basic mechanisms behind these behaviors have not been studied, it is predicted that their investigation will disclose routes to individually customized chalcogenide gas sensors. Additionally, for usage in commercial gas sensors, additional efforts must be made to enhance long-term stability and repeatability.

4.3.2 Carbon oxides

4.3.2.1 *Overview of CO_x*

Carbon oxides (CO_x) sometimes referred to as oxocarbons, are a family of organic molecules composed entirely of carbon and oxygen. Carbon monoxide (CO) and carbon dioxide (CO_2) are the most fundamental oxocarbons (Fig. 4.3). Numerous additional stable and metastable carbon oxides are known but are seen seldom.

CO is the simplest CO_x. CO is colorless, odorless, and somewhat lighter than air gas. CO is composed of one carbon atom and one oxygen atom. Carbon and oxygen atoms are separated by 112.8 pm, consistent with the presence of a triple bond. CO has the strongest chemical bond known, with a bond dissociation energy of 1072 kJ/mol. CO has three resonance configurations, however, the structure with the triple bond is the most accurate representation of the molecule's true electron density distribution. In coordination complexes, the CO ligand is referred to as carbonyl. CO is produced as a byproduct of incomplete combustion of fossil fuels such as gasoline, automated vehicles, aircraft, the wood used

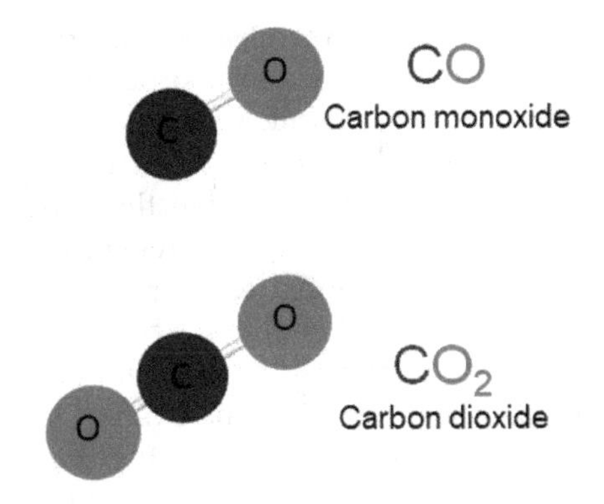

FIGURE 4.3 Carbon oxides (CO_x).

in boilers, natural gas emissions, coal, mines, gas fires, industrial waste, sewage leaks, solid fuel appliances, water heaters, and open fires, as well as other natural processes and substances on the Earth's surface. CO production is temperature and oxygen-dependent in this process. While thermal combustion is the most common source of CO, the gas is produced and released in significant quantities by a variety of environmental and natural sources.

When CO interacts with sunlight, O_3 is produced, which is thought to be detrimental to plants, animals, and the human respiratory system due to its strong oxidizing activity. Interrupting the body's regular oxygen supply puts the brain, heart, and other important organs at risk. Thus, early detection and notification of gas leaks are critical for safeguarding humans and work environments from CO infiltration. Decades of study have been devoted to the creation of optical, semiconductor, and electrochemical CO sensors. These sensors have a high sensitivity and are capable of detecting CO [50] (Zhou et al., 2018). Many typical CO sensors run at elevated temperatures (above 100°C), which consumes a lot of energy. As a result, a simple, efficient, and cost-effective sensor capable of measuring CO at elevated levels is required.

Carbon dioxide (chemical symbol CO_2) is a colorless, acidic gas with a density approximately 53% more than that of dry air CO_2molecules are formed via a covalent double bond between a carbon atom and two oxygen atoms. It occurs naturally as a trace gas in the Earth's atmosphere. Volcanoes, hot springs, and geysers are all-natural sources, and it is liberated from carbonate rocks by dissolving in water and acids. Since 1850, human activities have increased atmospheric CO_2 concentrations by almost 49% over pre-industrial levels. This exceeds what occurred naturally during 20,000 years (from the Last Glacial Maximum to 1850, from 185 ppm to 280 ppm). CO_2 is primarily an issue because it exacerbates the greenhouse effect. CO_2 and other pollutants together referred to as greenhouse gases, are released into the atmosphere as a result of both human activity and natural processes. These gases condense in the atmosphere and generate an insulating effect that wraps the globe in a blanket of warmth which is termed the greenhouse effect. They allow the sun's rays to enter, but when they bounce off the globe and back into space, the greenhouse gases reflect them to Earth. This process culminates in a rise in global temperature and a complicated series of events referred to as global climate change. CO_2 is monitored largely by a global network of atmospheric monitoring stations, such as the NOAA Mauna Loa Observatory, which offers single-site observations of greenhouse gas concentrations at the Earth's surface [51]. Satellite-based sounding devices augment these measurements. These measuring methodologies offer several advantages, including worldwide coverage and identification of several species. However, in situ monitoring with higher precision, shorter reaction time, and greater geographic resolution is necessary to provide a more thorough knowledge of climate change. Thus, to bring CO_x atmospheric concentrations down to tolerable levels, the need for dependable CO_x monitoring equipment such as sensors or pit gas concentration probes is critical.

4.3.2.2 *Chalcogenide-based gas sensor studies for* CO_x

Starecki et al., [52] created an all-optical infrared fiber sensor for insitu $\mathbf{CO_2}$ monitoring. A Dy3 + doped GaGeSbS fluorescent chalcogenide fiber serves as the 4.3 m mid-IR source testing the sensor's gas absorption. Following CO2's partial absorption at 4.3 m, a wavelength conversion from 4.3 m to 808 nm is achieved via excited state absorption processes in Er3 + doped GaGeSbS bulk glasses or fibers. This wavelength conversion enables the 808 nm transformed signal to be transported via silica fibers. This all-optical sensor has a sensitivity of a few hundred parts per million and can generally be deployed beyond a kilometer range, making it ideal for field operations. The photon conversion method employed in this gas sensor might be utilized to detect infrared radiations in general by employing visible or near-IR detectors rather than mid-infrared detectors.

Gutierrez-Arroyo et al., [31] developed a general model of an evanescent optical field sensor capable of detecting gases and compounds dissolved in water in the MIR. Ridge waveguides are presented as transducers, and their optimal dimensions were calculated using the Effective Index Method to offer the greatest evanescent field factor for monomodal propagation. The ideal length waveguide for any value of intrinsic loss propagation has been determined using the device's sensitivity. In gas detection for CO_2 at = 4.3 m, a numerical example has been presented utilizing a ridge waveguide base on (Ge-Sb-Se) chalcogenide glasses.

The study of carbon oxide-sensing properties of tellurium-based alloys is crucial for monitoring combustion exhausts. Several metal oxides have been proposed to help achieve this goal, including SnO_2, ZnO_2 and alkali metal carbonates, and copper [31,53]. However, these sensors are underutilized because of their poor selectivity and high operating temperatures (up to 500°C). As a result, it's vital to assess how well tellurium-based films detect CO. Concerning this, investigations were conducted by Tsiulyanu et al., [8] on the effects of propylamine ($C_3H_7NH_2$) and carbon oxide on the electrical conductivity of tellurium-based thin films. It was demonstrated that the absorption of propylamine vapor resulted in a reversible rise in the resistance of the layer, whereas the absorption of CO vapor resulted in a decrease in the resistance of the layer. The concentration of the gas influenced the sensitivity, as well as the response and recovery times.

4.3.3 Methane (CH_4)

4.3.3.1 Overview of CH_4

Methane is a chemical substance with the molecular symbol CH_4 attached to it (one atom of carbon and four atoms of hydrogen) (Fig. 4.4). It is a group 14 hydride, the most basic alkane, and the fundamental component of natural gas. CH_4 is an economically attractive fuel source because of its relative abundance on Earth; yet, collecting and storing it offers technological challenges due to its gaseous state at common temperature and pressure settings.

CH_4 may be found naturally underground and beneath the seafloor, where it is formed by geological and biological processes, and it is also found in the atmosphere. CH_4 clathrates, which are found under the ocean's surface, contain the world's largest known reservoir of CH_4. When it reaches the surface of the earth and enters the atmosphere, it is referred to as atmospheric CH_4. Since 1750, the amount of CH_4 in the atmosphere has increased by around 150%, accounting for approximately 20% of the globally mixed greenhouse gases [51]. Throughout the coal, natural gas, and oil sectors, CH_4 is both generated and delivered.

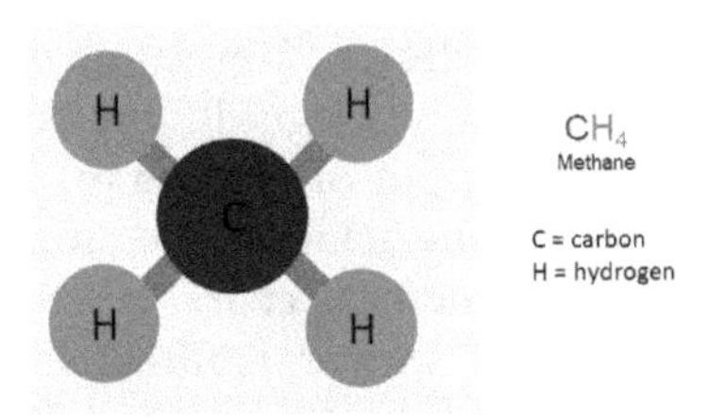

FIGURE 4.4 Methane (CH_4).

Additional sources of CH_4 emissions include cattle and other agricultural operations, land use, and the decomposition of organic material in municipal solid waste dumps, all of which contribute to global warming.

Methane has a heat of combustion of 55.5 MJ/kg and is a greenhouse gas. There are several steps involved in methane combustion, which are as follows:

$$CH_4 + 2O_2 \rightarrow CO_2 + 2H_2O \text{ (at normal circumstances, } H = 891 \text{ kJ/mol)}$$

Unlike CO_2, however, atmospheric methane is more difficult to quantify, sometimes requiring offline laboratory analysis utilizing a gas chromatograph and a flame ionization detector [54]. Quantification of methane is critical for analyzing the carbon cycle, as vertical profiling and better resolution give additional information on major methane sources and sinks.

The capacity to ascertain the chemical composition of or the presence of a chemical in liquid and gaseous mixtures such as CH_4 is critical in a wide variety of applications, including industrial process monitoring, environmental monitoring, forensics, medicine, and biology. Optical absorption spectroscopy is a frequently used technique for determining chemical compositions. It is generally conducted in the infrared region of the electromagnetic spectrum, where many compounds exhibit distinctive absorption spectra. The existing apparatus for measuring these spectra employs large and expensive free-space optics, whereas integrated photonics promises to accomplish the same analysis using substantially smaller chips that can be mass manufactured using processes similar to those employed in the electronics industry. To improve selectivity and sensitivity, these sensing components' operating wavelength can be extended to the mid-infrared (mid-IR), where distinctive spectroscopic properties reside. Various material systems have been used to make this possible. These include silicon, germanium, silicon-germanium, silicon nitride, semiconductors, and chalcogenides [55].

4.3.3.2 *Chalcogenide-based gas sensor studies for CH_4*

In a study, [5] discussed the design, fabrication, and analysis of a suspended slot chalcogenide glass waveguide gas sensor for methane as the target fas. Within a bandwidth of 3.0–4.4 m, a power confinement factor of greater than 90% may be achieved by rationally tuning the suggested waveguide sensor's structural characteristics. It was found the suggested waveguide sensor's power confinement factor, sensitivity, and limit of detection were 93.81%, 0.4578, and 18.17 ppm at a wavelength of 3.67 m, respectively. If the working wavelength is set to

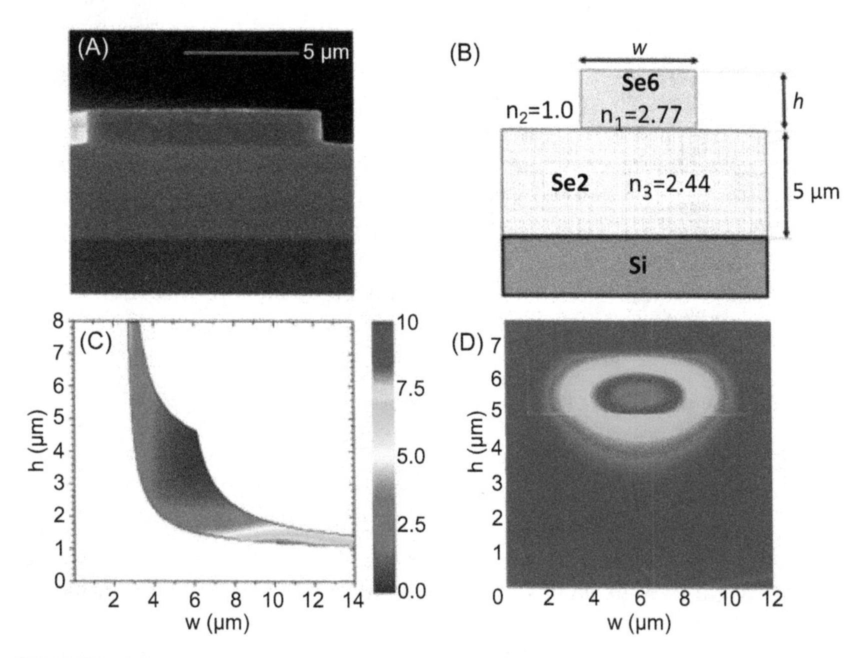

FIGURE 4.5 Schematic of a cross-section of chalcogenide ridge waveguide showing the dimensions and refractive index of the different layers [31].

3.291 m, the power confinement factor, sensitivity, and limit of detection are correspondingly 91.98%, 7.151%, and 1.139 ppm.

Gutierrez and coworkers [31] developed, manufactured, and optically evaluated a selenide integrated platform operating in the mid-infrared to detect CH_4. RF magnetron sputtering was used to deposit multilayered Ge-Sb-Se structures (Fig. 4.5). Ridge waveguides were produced as Y-junction, spiral, and S-shape waveguides using i-line photolithography and fluorine-based reactive ion etching. At 7.7 m, optical near-field imaging was used to study single-mode optical propagation, and optical propagation losses of 2.5 dB/cm were reported. Limits of detection of 14.2 parts per million for methane were detected utilizing this technology as an evanescent field sensor. Thus, these technical, practical, and theoretical discoveries provide a first step toward developing an integrated optical sensor working in the mid-infrared wavelength region. In another research, [31] employed a ridge Ge-Sb-Se chalcogenide glass waveguide sensor with an 8% PCF to detect CH4 at 3.31 m and 7.66 m, respectively.

Su et al., [55] developed a mid-infrared absorption spectroscopic sensing of CH_4 using an integrated chalcogenide glass spiral waveguide with an on-chip detector. Along with being a vital step toward a fully integrated chip-scale sensing system, waveguide integrated detectors provide lower noise as a result of their smaller size and higher speed. The sensor incorporated components of spiral chalcogenide glass waveguide sensors and PbTe detectors with waveguide integration.

The sensor's detection limit for methane gas was determined to be 1% by volume in an experiment utilizing a noise bandwidth of 0.078 Hz and the 3-r criteria using methane as a model gas. During measurements, the sensor shows no sign of drift, proving its high level of stability. The manufacturing process could be improved further, and the laser power variations should be normalized, resulting in a maximum sensitivity of 330 ppmv.

4.3.4 Hydrogen sulfide (H_2S)

4.3.4.1 Overview of H_2S

Hydrogen sulfide also known as H_2S (Fig. 4.6) is a gas that is heavier than air, toxic, extremely corrosive, and colorless. It has a malodor similar to that of rotten eggs. H_2S gas is frequently formed during the breakdown of organic materials, such as animal and human waste, in septic or bacterial sewage systems [56].

H_2S has a density somewhat greater than that of air. A combination of H_2S and air is explosive. H_2S burns with a blue flame in the presence of oxygen to produce sulfur dioxide (SO_2) and water. In general, H_2S functions as a reducing agent, particularly when combined with a base to create SH-[57] (Patnaik, 2007). SO_2 dioxide interacts with H_2S at elevated temperatures or in the presence of catalysts to generate elemental sulfur and water.

H_2S is utilized or manufactured in a variety of industries, including oil and gas refining; mining; tanning; pulp and paper processing, and rayon manufacturing. Natural sources of H_2S include sewers, manure pits, well water, oil and gas wells, and volcanoes. Since H_2S is heavier than air, it can accumulate in low-lying and confined places such as manholes, sewers, and underground telephone vaults [58]. Due to its presence, working in tight places may be quite risky. Additionally, it is generated through the decomposition of xanthene in the mining sector. H_2S is a precursor to the formation of sulfide and sulfuric acid.

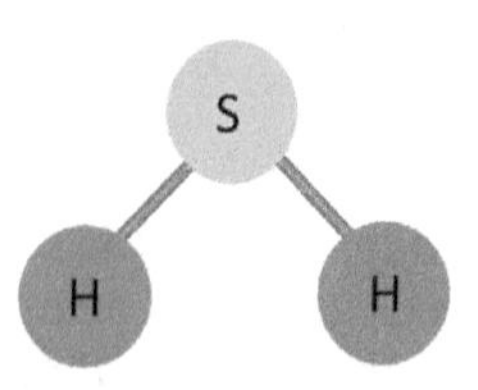

FIGURE 4.6 Hydrogen sulfide (H_2S).

H_2S is a very poisonous, highly flammable, and corrosive gas. It may occur naturally or may be discharged into the environment as a result of human activity. Human exposure to high levels of H_2S has a variety of adverse consequences. On the other hand, H_2S is created in vivo and serves as the third physiological gas transmitter with a variety of physiological functions. As a result, it is critical to creating sensitive, selective, and simple-to-use sensors to monitor H_2S in both its hazardous and physiological gas transmitter roles.

H_2S detection and monitoring techniques differ by sector and by the state, province, or nation in which a business operates. Numerous types of gas monitoring devices, in one form or another, are capable of continually collecting data on gas concentrations from a fixed place inside the monitored region. In certain circumstances, data logging is included as a standard feature of monitoring equipment. In other circumstances, the detector equipment might be linked to a personal computer or networked computer running data collection and analysis software. Gas chromatography-based technologies are frequently utilized as trustworthy procedures for the detection and monitoring of H_2S. However, this process has several disadvantages, including the fact that it requires many stages, is complicated, and introduces bias due to the reactivity of H_2S and its loss by sorption. As a result, numerous types of sensors for monitoring H_2S in the environment have been developed, including optical, electrochemical, conducting polymers, metal oxides, and nanosensors [59] (El-Shaheny et al.,) The majority of H_2S gas sensors are based on colorimetry and spectroscopic (absorption and fluorescence) principles. Future research should focus on developing more sensitive, robust, and stable sensors for H_2S and other critical trace contaminants that perform consistently under adverse environmental circumstances. Since the majority of these sensors have been validated in laboratory circumstances, they must be verified more thoroughly to withstand harsh environmental conditions. After all, one of the primary concerns in the actual implementation of H_2S -gas sensors would be cost-effectiveness.

4.3.4.2 *Chalcogenide-based gas sensor studies for H_2S*

Tellurium thin films have been discovered to be sensitive to a variety of reducing/oxidizing gases, including NO_2, CO, and NH_3 [48]. The films were deposited on single crystal and polycrystalline alumina substrates in a study by Sen et al. [34], and the interaction of H_2S with the films was examined. Thermal evaporation was used to create the films on alumina substrates at a temperature of 373K. The films were found to be sensitive to 0.1 ppm H_2S at ambient temperature. The reaction was repeatable, and the films remained stable during 3 months of operation. The sensors' detection process was explored, and it was discovered that H_2S reduced the quantity of oxygen adsorbed on the Te film surface, increasing resistance.

4.4 Summary

Chalcogenide compounds including Te, Ge, Se, As, and Mg are attractive options for future gas sensing technology. The sensitivity and selectivity of these sensors are determined by the composition of the sensing layer. In such layers, doping effects dominate over overall conductivity, resulting in improved signal-to-noise ratios in gas sensing applications, particularly when thin films are utilized. This opens new avenues for optimizing sensing qualities, including selectivity for certain gases and long-term stability under various conditions. In light of the large range of materials available in this glass family, sensor sensitivity may be tailored more precisely for different gas sensing applications. Existing technologies would be enhanced while new ones would be enabled if high-quality, low-loss, high-strength single-mode and multimode materials were available. The future of chalcogenide gas sensors looks to be promising.

References

[1] H.Y. Li, S.N. Zhao, S.Q. Zang, J. Li, Functional metal–organic frameworks as effective sensors of gases and volatile compounds, Chem. Soc. Rev. 49 (17) (2020) 6364–6401. Available from: https://doi.org/10.1039/C9CS00778D.

[2] E. Oleneva, T. Kuchmenko, E. Drozdova, A. Legin, D. Kirsanov, Identification of plastic toys contaminated with volatile organic compounds using QCM gas sensor array, Talanta 211 (2020) 120701. Available from: https://doi.org/10.1016/J.TALANTA.2019.120701.

[3] G. Jiang, T. Cumberland, X. Fu, M. Li, J. Zhang, P. Xu, et al., Highly stable low-cost electrochemical gas sensor with an alcohol-tolerant N, S-codoped non-precious metal catalyst air cathode, ACS Sens. 6 (3) (2021) 752–763. Available from: https://doi.org/10.1021/ACSSENSORS.0C01466/SUPPL_FILE/SE0C01466_SI_001.PDF.

[4] Y. Bao, P. Xu, S. Cai, H. Yu, X. Li, Detection of volatile-organic-compounds (VOCs) in solution using cantilever-based gas sensors, Talanta 182 (2018) 148–155. Available from: https://doi.org/10.1016/J.TALANTA.2018.01.086.

[5] J. Wang, H. Shen, Y. Xia, S. Komarneni, Light-activated room-temperature gas sensors based on metal oxide nanostructures: a review on recent advances, Ceram. Int. 47 (6) (2021) 7353–7368. Available from: https://doi.org/10.1016/J.CERAMINT.2020.11.187.

[6] H. Baltes, W. Göpel, J. Hesse, Sensors, update 1, Sensors 1 (3) (1996) 295. Retrieved from https://ui.adsabs.harvard.edu/abs/1996sens...0.1....B/abstract.

[7] L. Szaro, The effect of gaseous atmospheres on tellurium thin films, Thin Solid. Films 139 (1) (1986) 9–13. Available from: https://doi.org/10.1016/0040-6090(86)90042-8.

[8] D. Tsiulyanu, S. Marian, V. Miron, H.D. Liess, High sensitive tellurium based NO_2 gas sensor, Sens. Actuators B: Chem. 73 (1) (2001) 35–39. Available from: https://doi.org/10.1016/S0925-4005(00)00659-6.

[9] S. Mishra, P. Jaiswal, P. Lohia, D.K. Dwivedi, Chalcogenide glasses for sensor application: a review, in: Proceedings of the 2018 5th IEEE Uttar Pradesh Section International Conference on Electrical, Electronics and Computer Engineering, UPCON 2018. https://doi.org/10.1109/UPCON.2018.8596828, 2018.

[10] J. Wüsten, K. Potje-Kamloth, Chalcogenides for thin film NO sensors, Undefined 145 (1) (2010) 216–224. Available from: https://doi.org/10.1016/J.SNB.2009.11.058.

[11] C. Pénard-Morand, I. Annesi-Maesano, Air pollution: from sources of emissions to health effects, Breathe 1 (2) (2004) 108–119. Available from: https://doi.org/10.1183/18106838.0102.108.

[12] WHO, A. Quality and Health, Air pollution, in: WHO. Retrieved from https://www.who.int/health-topics/air-pollution#tab = tab_1, 2021.

[13] P.S. Myers, D. Foster, Vehicles and their powerplants: energy use and efficiency, In: Encyclopedia of Energy, Elsevier, (2004), pp. 347–362. Available from: https://doi.org/10.1016/B0-12-176480-X/00175-3.

[14] L. Tan, Y. Guo, Z. Liu, P. Feng, Z. Li, An investigation on the catalytic characteristic of NOx reduction in SCR systems, J. Taiwan. Inst. Chem. Eng. 99 (2019) 53–59. Available from: https://doi.org/10.1016/J.JTICE.2019.02.020.

[15] L. Cao, L. Chen, X. Wu, R. Ran, T. Xu, Z. Chen, et al., TRA and DRIFTS studies of the fast SCR reaction over CeO_2/TiO_2 catalyst at low temperatures, Appl. Catal. A: Gen. 557 (2018) 46–54. Available from: https://doi.org/10.1016/J.APCATA.2018.03.012.

[16] C. Chen, Y. Cao, S. Liu, J. Chen, W. Jia, Review on the latest developments in modified vanadium-titanium-based SCR catalysts, Chin. J. Catal. 39 (8) (2018) 1347–1365. Available from: https://doi.org/10.1016/S1872-2067(18)63090-6.

[17] N. Udhayakumar, S. Ramesh Babu, R. Bharathwaaj, R. Sathyamurthy, An experimental study on emission characteristics in compression ignition engine with silver and zinc coated catalytic converter, Mater. Today Proc. 47 (2021) 4959–4964. Available from: https://doi.org/10.1016/J.MATPR.2021.04.314.

[18] I. Yildiz, H. Caliskan, K. Mori, Effects of cordierite particulate filters on diesel engine exhaust emissions in terms of pollution prevention approaches for better environmental management, J. Environ. Manag. 293 (2021) 112873. Available from: https://doi.org/10.1016/J.JENVMAN.2021.112873.

[19] A. Zakery, S.R. Elliott, Optical nonlinearities in chalcogenide glasses and their applications, Springer Ser. Optical Sci. 135 (2007). Available from: https://doi.org/10.1007/978-3-540-71068-4.

[20] R.K. Jha, N. Bhat, Recent progress in chemiresistive gas sensing technology based on molybdenum and tungsten chalcogenide nanostructures, Adv. Mater. Interfaces 7 (7) (2020) 1901992. Available from: https://doi.org/10.1002/ADMI.201901992.

[21] D. Tsiulyanu, S. Marian, H. Liess, I. Eisele, Chalcogenide based gas sensors, J. Optoelectron. Adv. Mater. 5 (5) (2003) 1349–1354.

[22] H.J. Whitfield, Crystal structure of the β-form of tetra-arsenic trisulphide, J. Chem. Society Dalton Trans. 1 (17) (1973) 1737–1738. Available from: https://doi.org/10.1039/DT9730001737.

[23] S. Marian, K. Potje-Kamloth, D. Tsiulyanu, H.D. Liess, Dimorphite based gas sensitive thin films, Thin Solid. Films 359 (1) (2000) 108–112. Available from: https://doi.org/10.1016/S0040-6090(99)00707-5.

[24] D. Tsiulyanu, M. Ciobanu, Room temperature a.c. operating gas sensors based on quaternary chalcogenides, Sens. Actuators B Chem. 223 (2016) 95–100. Available from: https://doi.org/10.1016/J.SNB.2015.09.038.

[25] D. Tsiulyanu, G. Golban, E. Kolomeyko, O. Melnic, Photoconductivity and optical absorption of dimorphite thin films, Phys. Status Solidi (B) 197 (1) (1996) 61–64. Available from: https://doi.org/10.1002/PSSB.2221970110.

[26] S.K. Pandey, J. Singh, Nitrogen dioxide: risk assessment, environmental, and health hazard, Hazard. Gases (2021) 273–288. Available from: https://doi.org/10.1016/B978-0-323-89857-7.00001-3.

[27] D.L.H. Williams, Reagents effecting nitrosation, In: Nitrosation Reactions and the Chemistry of Nitric Oxide, Elsevier Science, 2004, pp. 1–34. Available from: https://doi.org/10.1016/B978-044451721-0/50002-5.

[28] F. Liu, S. Beirle, Q. Zhang, S. Dörner, K. He, T. Wagner, NOx lifetimes and emissions of cities and power plants in polluted background estimated by satellite observations, Atmos. Chem. Phys. 16 (8) (2016) 5283–5298. Available from: https://doi.org/10.5194/acp-16-5283-2016.
[29] J.G.J. Olivier, A.F. Bouwman, K.W. Van Der Hoek, J.J.M. Berdowski, Global air emission inventories for anthropogenic sources of NOx, NH_3 and N_2O in 1990, Environ. Pollut. 102 (1) (1998) 135–148. Available from: https://doi.org/10.1016/S0269-7491(98)80026-2.
[30] D. Tsiulyanu, S. Marian, H.D. Liess, I. Eisele, Effect of annealing and temperature on the NO_2 sensing properties of tellurium based films, Sens. Actuators B: Chem. 100 (3) (2004) 380–386. Available from: https://doi.org/10.1016/J.SNB.2004.02.005.
[31] A. Gutierrez-Arroyo, E. Baudet, L. Bodiou, J. Lemaitre, I. Hardy, F. Faijan, et al., Optical characterization at 7.7 μm of an integrated platform based on chalcogenide waveguides for sensing applications in the mid-infrared, Opt. Express 24 (20) (2016) 23109. Available from: https://doi.org/10.1364/OE.24.023109.
[32] D. Tsiulyanu, Tellurium thin films in sensor technology, Nanotechnological Basis for Advanced Sensors, NATO Science for Peace and Security Series B: Physics and Biophysics (2011) 363–380. Available from: https://doi.org/10.1007/978-94-007-0903-4_38.
[33] K. Okuyama, H. Kudo, Improved tellurium films by partially ionized vapor deposition as the semiconductor layer of a TFT and a hydrogen sensor, Japanese J. Appl. Phys. 28 (5R) (1989) 770–775. Available from: https://doi.org/10.1143/JJAP.28.770/XML.
[34] S. Sen, V. Bhandarkar, K.P. Muthe, M. Roy, S.K. Deshpande, R.C. Aiyer, et al., Highly sensitive hydrogen sulphide sensors operable at room temperature, Sens. Actuators B Chem. 115 (1) (2006) 270–275. Available from: https://doi.org/10.1016/J.SNB.2005.09.013.
[35] D. Tsiulyanu, S. Marian, H.D. Liess, Sensing properties of tellurium based thin films to propylamine and carbon oxide, Sens. Actuators B: Chem. 85 (3) (2002) 232–238. Available from: https://doi.org/10.1016/S0925-4005(02)00113-2.
[36] D. Tsiulyanu, A. Tsiulyanu, H.D. Liess, I. Eisele, Characterization of tellurium-based films for NO_2 detection, Thin Solid. Films 485 (1–2) (2005) 252–256. Available from: https://doi.org/10.1016/J.TSF.2005.03.045.
[37] L.A. Ba, M. Döring, V. Jamier, C. Jacob, Tellurium: an element with great biological potency and potential, Org. Biomol. Chem. 8 (19) (2010) 4203–4216. Available from: https://doi.org/10.1039/C0OB00086H.
[38] A.A. Wilhelm, C. Boussard-Plédel, Q. Coulombier, J. Lucas, B. Bureau, P. Lucas, Development of far-infrared-transmitting Te based glasses suitable for carbon dioxide detection and space optics, Adv. Mater. 19 (22) (2007) 3796–3800. Available from: https://doi.org/10.1002/ADMA.200700823.
[39] Z. Liu, T. Yamazaki, Y. Shen, T. Kikuta, N. Nakatani, T. Kawabata, Room temperature gas sensing of p-type TeO_2 nanowires, Appl. Phys. Lett. 90 (17) (2007) 173119. Available from: https://doi.org/10.1063/1.2732818.
[40] N. Dewan, S.P. Singh, K. Sreenivas, V. Gupta, Influence of temperature stability on the sensing properties of SAW NOx sensor, Sens. Actuators B Chem. 124 (2) (2007) 329–335. Available from: https://doi.org/10.1016/J.SNB.2006.12.036.
[41] T. Siciliano, M. Di Giulio, M. Tepore, E. Filippo, G. Micocci, A. Tepore, Tellurium sputtered thin films as NO_2 gas sensors, Sens. Actuators B Chem. 135 (1) (2008) 250–254. Available from: https://doi.org/10.1016/J.SNB.2008.08.018.
[42] A. Sharma, M. Tomar, V. Gupta, A low temperature operated NO_2 gas sensor based on TeO_2/SnO_2 p–n heterointerface, Sens. Actuators B Chem. 176 (2013) 875–883. Available from: https://doi.org/10.1016/J.SNB.2012.09.029.

[43] L. Guan, S. Wang, W. Gu, J. Zhuang, H. Jin, W. Zhang, et al., Ultrasensitive room-temperature detection of NO_2 with tellurium nanotube based chemiresistive sensor, Sens. Actuators B Chem. 196 (2014) 321–327. Available from: https://doi.org/10.1016/J.SNB.2014.02.014.
[44] M. Zhang, H.C. Su, Y. Rheem, C.M. Hangarter, N.V. Myung, A rapid room-temperature NO_2 sensor based on tellurium–SWNT hybrid nanostructures, J. Phys. Chem. C. 116 (37) (2012) 20067–20074. Available from: https://doi.org/10.1021/JP305393C.
[45] J. Park, Y.A. Soh, G. Aeppli, X. Feng, Y. Ou, K. He, et al., Crystallinity of tellurium capping and epitaxy of ferromagnetic topological insulator films on SrTiO3, Sci. Rep. 5 (1) (2015) 1–9. Available from: https://doi.org/10.1038/srep11595.
[46] Y.C. Her, B.Y. Yeh, S.L. Huang, Vapor-solid growth of p-Te/n-SnO2 hierarchical heterostructures and their enhanced room-temperature gas sensing properties, ACS Appl. Mater. Interfaces 6 (12) (2014) 9150–9159. Available from: https://doi.org/10.1021/AM5012518/SUPPL_FILE/AM5012518_SI_001.PDF.
[47] O. Mocreac, Effect of deposition rate and substrate microstructure on gas sensitivity of Te thin films, Zastita Mater. 60 (4) (2019) 379–384.
[48] Dumitru Tsiulyanu, Gas-sensing features of nanostructured tellurium thin films, Belistein J. Nanotechnol. 11 (2020) 1010–1018.
[49] Tiziana Siciliano, E. Filippo, A. Genga, G. Micocci, M. Siciliano, A. Tepore, Single-crystalline Te microtubes: synthesis and NO_2 gas sensor application, Sens. Actuators B Chem. 142 (1) (2009) 185–190. Available from: https://doi.org/10.1016/J.SNB.2009.07.050.
[50] M. Shojaee, S. Nasresfahani, M.H. Sheikhi, Hydrothermally synthesized Pd-loaded SnO_2/partially reduced graphene oxide nanocomposite for effective detection of carbon monoxide at room temperature, Sens. Actuators, B: Chem. 254 (2018) 457–467. Available from: https://doi.org/10.1016/J.SNB.2017.07.083.
[51] S. Solomon, D. Qin, M. Manning, Z. Chen, M. Marquis, K.B. Averyt, et al., Climate change 2007: the physical science basis contribution of working group I to the fourth assessment report of the Intergovernmental Panel on Climate Change, in: Lic.Wisc. Edu. Retrieved from http://www.ipcc.ch/publications_and_data/publications_ipcc_fourth_assessment_report_wg..., 2007.
[52] F. Starecki, A. Braud, J.L. Doualan, J. Ari, C. Boussard-Plédel, K. Michel, et al., All-optical carbon dioxide remote sensing using rare earth doped chalcogenide fibers, Opt. Lasers Eng. 122 (2019) 328–334. Available from: https://doi.org/10.1016/J.OPTLASENG.2019.06.018.
[53] R.K. Sharma, P.C.H. Chan, Z. Tang, G. Yan, I.-M. Hsing, J.K.O. Sin, Sensitive, selective and stable tin dioxide thin-films for carbon monoxide and hydrogen sensing in integrated gas sensor array applications, Sens. Actuators B. Chem. 2 (72) (2001) 160–166. Retrieved from https://www.infona.pl//resource/bwmeta1.element.elsevier-32a1d543-de96–3687-a88e-6ddf451b4d80.
[54] E.J. Dlugokencky, L.P. Steele, P.M. Lang, K.A. Masarie, The growth rate and distribution of atmospheric methane, J. Geophys. Res. Atmos 99 (D8) (1994) 17021–17043. Available from: https://doi.org/10.1029/94JD01245.
[55] P. Su, Z. Han, D. Kita, P. Becla, H. Lin, S. Deckoff-Jones, et al., Monolithic on-chip mid-IR methane gas sensor with waveguide-integrated detector, Appl. Phys. Lett. 114 (5) (2019) 051103. Available from: https://doi.org/10.1063/1.5053599.
[56] E.M. Bos, H. Van Goor, J.A. Joles, M. Whiteman, H.G.D. Leuvenink, Hydrogen sulfide: physiological properties and therapeutic potential in ischaemia, Br. J. Pharmacol. 172 (6) (2015) 1479–1493. Available from: https://doi.org/10.1111/BPH.12869.
[57] P. Patnaik, A comprehensive guide to the hazardous properties of chemical substances: third edition, in: A Comprehensive Guide to the Hazardous Properties of Chemical Substances. https://doi.org/10.1002/9780470134955, 2006.

[58] E. Vaiopoulou, P. Melidis, A. Aivasidis, Sulfide removal in wastewater from petrochemical industries by autotrophic denitrification, Water Res. 39 (17) (2005) 4101–4109. Available from: https://doi.org/10.1016/J.WATRES.2005.07.022.

[59] R. El-Shaheny, F. Belal, Y. El-Shabrawy, M. El-Maghrabey, Nanostructures-based sensing strategies for hydrogen sulfide, Trends Environ. Anal. Chem. 31 (2021) e00133. Available from: https://doi.org/10.1016/J.TEAC.2021.E00133.

CHAPTER

5

Chemical (bio)sensors

8.1 Drug

Let's start this section by asking a question, though there are blood and urine tests for the detection of drugs, why these traditional methods for drug detection are not useful anymore? Traditional drug detection is a complex process that needs a series of complicated and successive extracting and analyzing methods, for example, (1) extracting suspected drugs from biological specimens, and (2) analyzing drugs through various instruments. Moreover, the laboratory kits can only detect a single component of a drug, in a single test, therefore it has a low sensitivity. Furthermore, blood and urine testing have shortcomings in regard to the concerns of declining athletic performance and violating human rights as the tester, respectively [1].

Drug precursor detection using a portable device at a tiny amount is very important, particularly in laboratories [2]. Since many drugs are formulated using more than one chemical in a very low concentration, the sensing devices for detection must be highly selective, and sensitive.

Drug detection is important not only for laboratory recording but also in many other critical applications, such as:

- Screening suspects who have been apprehended.
- Searching drugs in suspected storage areas.
- Screening prisoners in or visitors via pass-through designated checkpoints.
- Screening large numbers of people from the general public.
- Screening hand-carried items at busy personnel checkpoints.
- Screening large numbers of vehicles at crowded checkpoints.
- Screening letters, packages, and other items that pass through a mailroom.

Metal Chalcogenide Biosensors
DOI: https://doi.org/10.1016/B978-0-323-85381-1.00001-5

Humans emit characteristic sign-of-life volatile organic compounds. The types and concentration ranges of specific volatiles are differed, which can detect via sensors. Gas and electrochemical sensors are developed to determine human presence within a few minutes of direct measurements in the field, however, recognition of trapped and/or hidden humans with a higher level of confidence is still required [3].

Over the last two decades ago, scientists have tried to develop the technology of mobile sensors. A sweat sensor is a very new sensing device that is not invasive and relatively free from human rights issues. In wearable sweat-based sensors, very small amounts of substances are discharged onto thin and flexible plastic such as the technology of printing press imprinting. The sensor contains helix-like microfluidic tubes, that can derive information on the rate at which the wearer is sweating.

Several projects worldwide have been defined, developed, and funded for the fabrication of drug sensing devices by complementing low-cost multiple techniques, that is, portable, affordable, and selective optical sensing devices for detection of drug-chemical-clusters of organic and/or organometallic substances such as amphetamine, ephedrine, safrole, gemcitabine, Bortezomib, acetic anhydride, etc.

Nanostructured metal chalcogenides are developed as promising electrode materials for drug detection. Metal chalcogenides have excellent electron transfer and good sensing behavior, much higher than metal oxides. These materials are widely utilized in energy and electrochemical sensor applications of drugs. For an instant, cerium-ruthenium sulfide nanostructures (CRS-NSs) are used for electrochemical sensing of trifluoperazine (TFPZ) [1]. TFPZ is a psychiatric medication that belongs to the class of drugs called phenothiazine antipsychotics [4]. In this study, the metal chalcogenide is prepared via hydrothermal technique, and shows almost a good electrocatalytic activity towards the electrochemical oxidation of TFPZ with a distinct amperometric response with the lowest limit of detection (LOD) of 0.322 nM ($S/N = 3$), high sensitivity 2.682 $\mu A/\mu M\ cm^2$ and lowest oxidation potential of +0.64 V (Ag/AgCl). Furthermore, the Ce–Ru–S NS displays excellent selectivity, good reproducibility, and long-term stability.

An imipramine (IMPR) sensor, based on metal chalcogenide-carbon composite materials, is fabricated by Yamuna et al. [5]. In this study, the antimony telluride–graphite nanofiber (Sb_2Te_3-GNF, hereafter SBT-GNF) composite is synthesized via a wet-chemistry method (hydrothermal) and used as a disposable sensor electrode with a screen-printed carbon electrode (SPCE). This device is utilized for the detection of IMPR by voltammetry. The results showed the SBT-GNF/SPCE has an acceptable linear range (0.01-51.8 μM), sand sensitivity (1.35 ± 0.1 $\mu A/\mu M\ cm^2$) with high selectivity, and high tolerance limit against potential interfering compounds in blood serum and urine samples.

In a similar vein, a group of researchers from Taiwan [6] reported a flower-like structured Ce–Ru–S nanostructure for the detection of antipsychotic drug trifluoperazine (TFPZ) in human urine. They synthesized the starting materials via one-pot hydrothermal synthesis. They expressed that the rich redox chemistry and synergistic activity of Ce–Ru–S NS can potentially enhance the electrochemical performance of TFPZ. The sensor exhibited a low oxidation potential, wide linear range, good selectivity, and precision against TFPZ.

In summary, and based on reported articles, it can be deduced that the nanostructured binary metal chalcogenide is considered a promising electrode material with excellent electron transfer and good sensing behavior, rather than metal oxides, for the detection of various functional drugs.

8.2 Trace elements

Chemical sensors can be divided into several types, such as optical, electrochemical, mass, magnetic, and thermal. Nanostructured materials consist of nanoparticles (NPs), nanorods, nanowires, thin films, or bulk materials made of nanoscale building blocks or composed of nanoscale structures with at least one dimension falling on a nanometer scale [7].

In-situ detection of trace elements in industrial effluents (heavy metals) or drinking water (amount of trace elements) is very important in terms of "environment" and "human health" [8]. Though sometimes the concentration of hazardous elements is naturally high in the soil in some areas, various human activities can also cause trace element pollution of the aquatic environment such as mining, coal and fuel combustion, and industrial processing [9]. Industrial processing is one of the most renowned activities ranging from chemical and metal production to petroleum and agriculture. The chemical industry produces various industrial chemicals, where raw chemicals into more than 70,000 different products. Toxic effluents from these industries can affect health and physical hazards to human health and the environment such as sickness, injury, and even death. The hazardous materials from chemical industries are in the form of gases or vapors, liquids, dust, and solids. The hazardous chemical industries are as:

- Metal and alloy industries: metal fabricators, steel producers, iron makers, and various related industries deal with some hazardous materials routinely. Coke production during steel manufacturing is one of the primary hazardous materials sources. Coke ovens emit ammonium compounds, naphthalene, crude light oil, coke dust, and sulfur. Inorganic arsenic is also found in coke oven emissions from

smelter processes. Arsenic and arsenic compounds are found in powder, crystalline, vitreous, or amorphous forms.
- Chloro-alkali industries: in which chlorine and caustic soda (sodium hydroxide) are produced. The method of production is generally via electrolytic decomposition of salt (sodium chloride). Since the 20th century, salt is used as the principal source of chlorine.
- Petroleum industries: is enormous with activities that extended globally, ranging from draining to refining. Crude oil refining processes generate significant amounts of sludge, which is a complex mixture of trace metals and many other organic compounds. These hazard elements can affect soil, water, air, and therefore human health [10].
- Agriculture: fertilizers, pesticides, and herbicides are three important components of agriculture hazardous materials.
- Effluents or sewage: the effluents from residential, commercial, and particularly industrial activities are known as the major hazardous chemicals containing various organic and inorganic hazardous materials.
- Nuclear activity: nuclear waste is particularly hazardous and hard to manage relative to other toxic industrial wastes. Radioactive materials are the main domains of hazardous chemicals from nuclear industries.

There are several crucial sources of elemental or heavy metal input ranging from atmospheric fallout to lithosphere (leaching/dumping), or directly into the aquatic environment. The latter include ground, surface, river, lakes, estuarine, seas, and oceans. Here, there are six hazardous sources of soluble metal ions: cadmium (Cd), copper (Cu), iron (Fe), lead (Pb), nickel (Ni), and zinc (Zn). Almost all the above elements can be found in trace or high concentrations in the effluents of as mentioned industries.

The magnitude (concentration) of trace element input, duration of input, physical and chemical form, and associated ligands plays a critical role in the impact of water pollution based on their availability, transport, and toxicity. Naturally, the concentration of these elements differs in fresh water, river water, and seawater. For example, the typical natural concentration of cadmium in fresh water and river water is around 0.02 ppb, while it is 0.10 ppb in seawater. In the same vein, the concentration of Fe in freshwater is reported to be around 500 ppb, while it is 40 and 2 ppb in river and seawater, respectively. Therefore, it is crucial to detect the concentration of these soluble elements in the water. pH, solubility, temperature, and the nature of other chemical species are the major factors in determining the impact of these elements. For example, some chemical forms (chemical species) of above mentioned trace elements in various water sources are $CdCl^{+}$, $Cu(OH)^{+}$, $[Fe(OH)_2]^{+}$, $PbCl^{+}$, $Zn(OH)^{+}$, etc.

Many trace metal ion contaminants entering the aquatic environment can have a dramatic effect on the bioavailability and toxicity of biological processes. In particular, bio-amplification by plankton, or biotransformation by bacteria in the water-sediment interface can strongly influence elemental toxicity throughout the remaining food chain. For example, lead undergoes biomethylation in the water-sediment interface, resulting in the production of more toxic species which are concentrated in shellfish or fish. During the past 50 years, there has been a rapid increase in the number of major traces of element-related water pollution incidents.

- Cadmium is highly toxic and has been implicated in some cases of poisoning through food. Minute quantities of cadmium are suspected of being responsible for adverse changes in the arteries of human kidneys. Cadmium also causes generalized cancers in laboratory animals and has been linked epidemiologically with certain human cancers. A cadmium concentration of 200 μg/L is toxic to certain fish. Cadmium may enter the water because of industrial discharges or the deterioration of galvanized pipes.
- Copper salts are used in water supply systems to control biological growths in reservoirs and distribution pipes and to catalyze the oxidation of manganese. Corrosion of copper-containing alloys in pipe fittings may introduce measurable amounts of copper into the water in a pipe system.
- In water samples, iron may occur in true solution, in a colloidal state that may be peptized by organic matter, in inorganic or organic iron complexes, or relatively coarse suspended particles. It may be either ferrous or ferric, suspended or dissolved. Iron in water can cause staining of laundry and porcelain. A bittersweet astringent taste is detectable by some persons at levels above 1 mg/L.
- Lead is a serious cumulative body poison. Natural waters seldom contain more than 5 μg/L, although much higher values have been reported. Lead in a water supply may come from industrial, mine, and smelter discharges, or the dissolution of old lead plumbing. Tap waters that are soft, acid, and not suitably treated may contain lead resulting from an attack on lead service pipes or solder pipe joints.
- Zinc is an essential and beneficial element in human growth. Concentrations above 5 mg/L can cause a bitter astringent taste and an opalescence in alkaline waters. Zinc most commonly enters the domestic water supply from the deterioration of galvanized iron and dezincification of brass. In such cases lead and cadmium may also be present because they are impurities of the zinc used in galvanizing. Zinc in water may also result from industrial waste pollution.

Some compounds of nickel are highly toxic and may be carcinogenic. Potential symptoms of overexposure are sensitization dermatitis, allergic asthma, and pneumonitis. It is used in nickel-plating, as a catalyst for hydrogenation reactions, and in stainless steels, heat, and corrosion-resistant alloys, and alloys for electronic and space applications.

For these reasons, it is important to be able to detect the presence of these metal ions and to quantify the amount present quickly and accurately. Further, removal methods are necessary for instances when the amount of these trace metal ions is dangerously high.

The sampling of water requires careful procedures and can introduce potential errors in measurement. Because most trace metal ions to be measured are at very low levels, sample contamination and analyte losses are potential problems. Special precautions are necessary for samples containing trace metal ions. Because many constituents may be present at concentrations of micrograms per liter, they may be totally or partially lost if proper sampling and preservation procedures are not followed. Cadmium, copper, iron, lead, and zinc are subject to lose by adsorption on, or ion exchange with, container walls, or by precipitation.

Typical methods of analysis include atomic absorption spectroscopy, inductively coupled plasma emission spectroscopy, and colorimetric methods. Water samples must be collected and transported to a lab, where they are then analyzed. Sample pretreatment including filtration and acid digestion is often necessary. This is time-consuming and further increases the chances of contamination and imprecision.

For these reasons, it would be advantageous to develop a method to detect and quantify cadmium, copper, iron, lead, nickel, and zinc ions in water and effluent samples that can be performed in situ, requiring no sample storage, transport, or pretreatment.

8.3 Adulteration in different fuels

Petroleum fuels are vulnerable to adulteration for profit margins which causes a loss in engine performance and environmental hazards. Detection of such impurities is difficult since the adulterants are usually those compounds that exist in the fuels. Therefore, efficient analytical methods are urgently required for monitoring "adulteration" [11]. Real-time detection of fuel adulteration is crucial to control many hazardous effects on the environment and living organisms. Various automotive users and fuel providers may mix kerosene in petrol and diesel. Adulterated fuel increases air pollution, as well as reduces the performance of vehicle engines [12]. Adulteration in fuel involves the mixing of oil products with low-grade products such as refined products or condensates with high-demand products.

Various types of adulteration can be classified as:

- Blending relatively small amounts of distillate fuels such as Kerosene.
- Blending variable amounts of gasoline boiling range hydrocarbons.
- Blending small amounts of waste industrial solvents such as lubricants.
- Blending kerosene into diesel.
- Blending small amounts of heavier fuel oils into diesel fuels

Detection of fuel adulteration is challenging as it may naturally exist in the compounds. Several physicochemical measurements are applied for controlling compositional variations of these fuels. All these conventional measurements are possessed several limitations to accuracy in discriminating between adulterated samples the unaltered ones. Therefore, mixing statistical designs along with data mining can be useful. Recently, the monitoring of this phenomenon is detected using sensing devices such as optical fiber sensors. These devices can be used at the distribution point, to prevent adulteration in a short time [13].

8.4 Summary

The state-of-the-art in chemical sensor technology and applications are urgently required in modern lives. Highly unified and portable wireless chemical sensor systems are widely fabricated for the detection of various materials including drugs, trace elements, and adulterations. The significance of petroleum product adulteration in most countries is critical owing to urbanization, population increase, development events, and lifestyle changes, which in turn lead to prevalent environmental pollution.

References

[1] A. Sangili, R. Sakthivel, S.M. Chen, Cost-effective single-step synthesis of flower-like cerium-ruthenium-sulfide for the determination of antipsychotic drug trifluoperazine in human urine samples, Anal. Chim. Acta. 1131 (2020) 35–44. Available from: https://doi.org/10.1016/J.ACA.2020.07.032.

[2] P. Sundaresan, A. Yamuna, S.M. Chen, Sonochemical synthesis of samarium tungstate nanoparticles for the electrochemical detection of nilutamide, Ultrason. Sonochem. 67 (2020) 105146. Available from: https://doi.org/10.1016/j.ultsonch.2020.105146.

[3] E.M. Bos, H. Van Goor, J.A. Joles, M. Whiteman, H.G.D. Leuvenink, Hydrogen sulfide: physiological properties and therapeutic potential in ischaemia, Br. J. Pharmacol. 172 (2015) 1479–1493. Available from: https://doi.org/10.1111/BPH.12869.

[4] L. de, O. Marques, B. Soares, M.S. de Lima, Trifluoperazine for schizophrenia, Cochrane Database Syst. Rev. (2004). Available from: https://doi.org/10.1002/14651858.CD003545.PUB2.

[5] A. Yamuna, T.W. Chen, S.M. Chen, A.M. Al-Mohaimeed, W.A. Al-onazi, M.S. Elshikh, Selective electrochemical detection of antidepressant drug imipramine in blood serum and urine samples using an antimony telluride-graphite nanofiber electrode, Microchim. Acta 188 (2021) 1–12. Available from: https://doi.org/10.1007/S00604-021-04722-3. 2021 1882.

[6] A. Sangili, R. Sakthivel, S.M. Chen, Cost-effective single-step synthesis of flower-like cerium-ruthenium-sulfide for the determination of antipsychotic drug trifluoperazine in human urine samples, Anal. Chim. Acta. 1131 (2020) 35–44. Available from: https://doi.org/10.1016/J.ACA.2020.07.032.

[7] T. Kenny, Chemical sensors, Sensor Technology Handbook, Newnes, 2005, pp. 181–191. Available from: https://doi.org/10.1016/B978-075067729-5/50047-1.

[8] S.I. Ohira, Y. Miki, T. Matsuzaki, N. Nakamura, Y. ki Sato, Y. Hirose, et al., A fiber optic sensor with a metal organic framework as a sensing material for trace levels of water in industrial gases, Anal. Chim. Acta. 886 (2015) 188–193. Available from: https://doi.org/10.1016/J.ACA.2015.05.045.

[9] L. Pujol, D. Evrard, K. Groenen-Serrano, M. Freyssinier, A. Ruffien-Cizsak, P. Gros, Electrochemical sensors and devices for heavy metals assay in water: the French groups' contribution, Front. Chem. 2 (2014) 19. Available from: https://doi.org/10.3389/FCHEM.2014.00019/BIBTEX.

[10] P.E. Rosenfeld, L.G.H. Feng, The petroleum industry, Risks of Hazardous Wastes, William Andrew Publishing, 2011, pp. 57–71. Available from: https://doi.org/10.1016/B978-1-4377-7842-7.00005-2.

[11] B.P. Vempatapu, P.K. Kanaujia, Monitoring petroleum fuel adulteration: a review of analytical methods, TrAC. Trends Anal. Chem. 92 (2017) 1–11. Available from: https://doi.org/10.1016/J.TRAC.2017.04.011.

[12] A.S. Fadairo, J. Ekoh-Chukwukalu, G.A. Adeyemi, O.G. Abolarin, I.M.F. Mkpaoro, A fast and cost-efficient method to detect ethanol as adulterant in gasoline, MethodsX 7 (2020) 100974. Available from: https://doi.org/10.1016/J.MEX.2020.100974.

[13] S. Dilip kumar, T.V. Sivasubramonia Pillai, Estimating fuel adulteration in automobiles using robust optical fiber sensors, Microprocess. Microsyst. 79 (2020) 103289. Available from: https://doi.org/10.1016/J.MICPRO.2020.103289.

CHAPTER

6

Environmental sensors

6.1 Introduction

Environmental monitoring is important for both human health and environmental protection. Despite advances in pollution control, pollution will continue to be generated as the human population grows, as industrial growth and energy use expands, and as the human population grows. As a result, the need for environmental monitoring remains as strong as ever. To improve the accuracy and cost-effectiveness of monitoring programs, continued advancements in the production, implementation, and automation of monitoring devices are needed. The need to produce more scientists and engineers with the skills and training needed to effectively create and operate monitoring devices and manage monitoring programs is also critical. Environmental monitoring is the act of observing and studying the natural world which is rooted in the scientific method. The usage of environmental sensors is very essential in environmental monitoring activities. Environmental monitoring sensors enable the collection and analysis of specific data to characterize and monitor the environment's quality.

Environmental sensors are linked objects that can provide a variety of data, including location, position, the individual's movements, and contextual elements, which can be compared to data collected by sensors embedded on or implanted in the individual, as well as the validation of alarms. The use of sensors to track the environment has grown in popularity over the last few decades. Environmental phenomena such as weather and earthquakes, volcanoes, air quality, irrigation systems, forests, and ecological systems can all be monitored. Environmental sensors, such as biological or chemical sensors, have recently become more accurate and lightweight, making their use in remote areas where human sampling is impractical, as well as for microscopic applications, very realistic. These sensors are

Metal Chalcogenide Biosensors
DOI: https://doi.org/10.1016/B978-0-323-85381-1.00009-X

commonly used to track infrastructure and the environment. Sensors that can measure parameters that define air quality and dissolved constituents and the existence of substances in water, groundwater, and soils are needed in most environmental monitoring applications. Commercially available sensors for environmental application monitoring are available but they are typically costly for large-scale distributed applications. The integrity of off-site laboratory analyses can be impeded during the sample selection, transportation, storage, and analysis processes, which can take several days or longer. Reliable, low-cost, long-term monitoring of environmental pollutants using sensors that can be controlled on-site or in situ is needed. Sensitivity, accuracy, precision, reversibility, speed, durability, reliability, simplicity, selectivity, affordability, and acceptability are among the main requirements for successful sensor-based long-term environmental application monitoring. This chapter reviews the emergence of sensors-based technology with particular emphasis on chalcogenides for environmental monitoring. This includes the state-of-the-art of this technology in the detection of heavy metals and pesticides in various environmental segments such as water, air, and soil. The development of sensors-based chalcogenides for extreme environmental conditions such as under high-temperature conditions and radiation sensing is also addressed.

6.2 Heavy metals

Metals, except for alkaline and alkaline earth metals, have a density of (d) >5 g/mL and an atomic number of >20. Metals make up less than 0.1% of the earth's crust. While the word "heavy metals" originally referred to elements with high cellular toxicity, it has now been expanded to include micronutrients that pose a health risk at high concentrations. They are defined as metals with relatively high densities, atomic weights, or atomic numbers that are classified as heavy metals. Lead (Pb), arsenic (As), mercury (Hg), cadmium (Cd), zinc (Zn), silver (Ag), copper (Cu), iron (Fe), chromium (Cr), nickel (Ni), palladium (Pd), and platinum (Pt) are a few examples. The most prevalent heavy metal pollutants in the environment are Cr, Mn, Ni, Cu, Zn, Cd, and Pb [1].

Human and anthropogenic causes, such as industrial discharge, vehicle emissions, and mining, release these metals (Fig. 6.1). These include activities from industries, mining, and agricultural processes, posing threats to both ecological and human health. Heavy metal-induced toxicity and carcinogenicity are complicated by several mechanisms, some of which are not well known. Each metal, on the other hand, is known to have distinct characteristics and physicochemical properties that contribute to its distinct toxicological mechanisms of action. Table 6.1 summarizes the sources, environmental effects, and human health risks

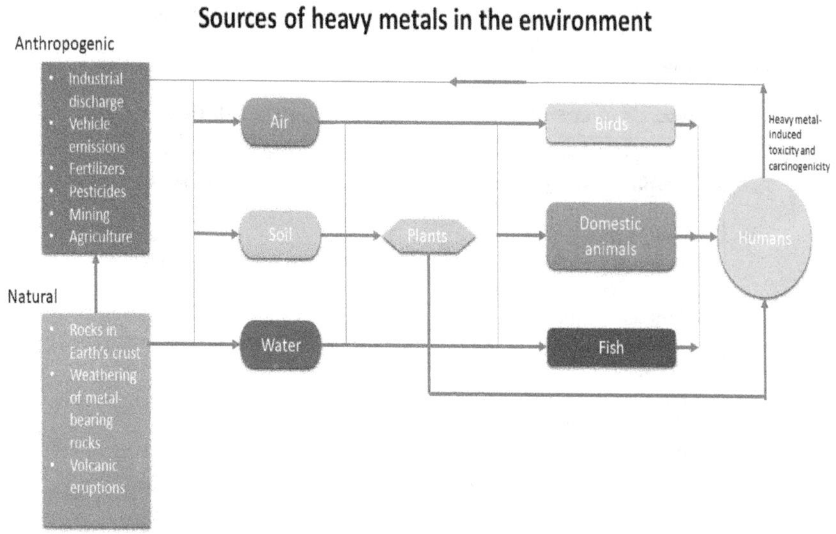

FIGURE 6.1 Sources of heavy metals in the environment.

of toxic heavy metals in terms of arsenic, cadmium, chromium, lead, and mercury. The information is adapted from Ref. [1].

Heavy metals, unlike organic contaminants, are nonbiodegradable and appear to accumulate in living organisms. Many of them are carcinogens. Long-term and constant exposure to heavy metals has been linked to many human health risks. Because of their pervasiveness and prevalence, heavy metal pollution has been identified as a major environmental concern. Since they are nondegradable and appear to bioaccumulate, effective methods that are functional, cost-effective, and quick to deploy in a variety of physical settings, for their removal from the environment must be developed.

6.2.1 Sensors-based chalcogenides for heavy metal detection

Many methods have been developed to detect heavy metals. Chemical adsorption, biosorption, electrolysis, flocculation, ion exchange, and membrane separation are some of the heavy metals removal methods that have been documented in the literature. [2], for example, summarized several studies on the identification of trace heavy metals using conducting polymers-based sensors. In another report by [3], the recent developments in electrochemical methods for trace-level ion in situ heavy metal sensors are discussed. Amongst the existing methods, as stated in Ref. [4], the most commonly used analytical techniques for detecting heavy metals are atom absorption spectroscopy (AAS), optical emission spectrometry with

TABLE 6.1 Sources, environmental effects, and human health risks of heavy metals in terms of arsenic, cadmium, chromium, lead, and mercury.

Heavy metal	Sources and environmental effects	Human health risks
Arsenic (As)	• Natural events such as volcanic eruptions and soil erosion, as well as anthropogenic activities, contribute to arsenic emissions in the environment. • Insecticides, herbicides, fungicides, algicides, sheep dips, wood preservatives, and dyestuffs are just a few of the arsenic-containing compounds that are manufactured industrially. • Arsenic is present in high concentrations in groundwater.	• Arsenic is absorbed by the mouth (ingestion), inhalation, dermal contact, and, to a lesser degree, parenteral administration. • Arsenic exposure has a detrimental effect on nearly every organ system, including the cardiovascular, dermatologic, and respiratory systems.
Cadmium (Cd)	• Cadmium compounds accumulate at the highest concentration in the environment in sedimentary rocks, and marine phosphates. • Cadmium's primary industrial applications include the manufacture of alloys, pigments, and batteries.	• The most common routes of exposure to cadmium are inhalation or cigarette smoke, as well as food consumption. • Chronic inhalation exposure to cadmium particulates is generally associated with pulmonary function, bone mineral density, and osteoporosis.
Chromium (Cr)	• Metal processing, tannery facilities, chromate production, stainless steel welding, and ferrochrome and chrome pigment production are the industries that contribute the most to chromium release. • Chromium enters and polluted various environmental segments (air, water, and soil) via a variety of natural and anthropogenic sources.	• The health risk associated with chromium exposure is determined by its oxidation state, which ranges from low toxicity in the metal form to high toxicity in the hexavalent form. • Although inhalation is the main route of human exposure to chromium and the lung is the primary target organ, substantial human exposure to chromium has also been identified via the skin. • occupational and environmental exposure to chromium-containing compounds has been linked to multiorgan toxicity, including renal injury, allergy and asthma, and respiratory cancer. • Exposure to chromium has been linked to dermatitis among construction workers.

(Continued)

TABLE 6.1 (Continued)

Heavy metal	Sources and environmental effects	Human health risks
Lead (Pb)	• Lead is found naturally in the atmosphere, and anthropogenic activities such as the burning of fossil fuels, mining, and manufacturing contribute to the release of high levels. • Lead in dust and soil contaminate houses and the environment	• Lead poisoning is caused by inhaling lead-contaminated dust particles or aerosols, as well as ingesting lead-contaminated food, water, and paints. • The kidney, liver, and other soft tissues absorb the most lead in the human body resulting in lead poisoning affecting the nervous system.
Mercury (Hg)	• Mercury is used in the electrical industry, dentistry, and a variety of industrial processes. • Mercury is a pervasive environmental contaminant and pollutant found in soil and water.	• Dental amalgams and fish consumption are the primary sources of chronic, low-level mercury exposure. • The kidneys, neurological tissue, and liver accumulate a significant portion of what is absorbed. Mercury in all its forms is toxic, causing gastrointestinal toxicity, neurotoxicity, and nephrotoxicity.

Adapted P.B. Tchounwou, C.G. Yedjou, A.K. Patlolla, D.J. Sutton, Heavy metal toxicity and the environment, Mol. Clin. Environ. Toxicol., pp. 133–164, 2012, https://doi.org/10.1007/978-3-7643-8340-4_6

inductively coupled plasma (ICP-OES), polarography, and ion chromatography (IC). However, the devices required are extremely complicated and require skilled operators. These methods do not allow for measurements to be taken directly on-site. Their high operational and equipment costs are also drawbacks. A simple experimental setup, therefore, becomes increasingly necessary. Chalcogenide glasses (ChG) have been demonstrated to be promising materials for use as sensitive membranes in ion-selective electrodes (ISEs) used in electrochemical sensors to detect heavy metal ions in solutions. Chalcogenide glasses are made up of the chalcogen elements sulfur (S), selenium (Se), and tellurium (Te), and other elements including germanium (Ge), arsenic (As), and antimony (Sb) are added to make stable glasses. Owing to their clarity in the infrared optical spectrum and their ability to be drawn into optical fibers, chalcogenide glasses are well-known materials. Numerous studies have been carried out since the 1970s pertaining to sensors-based chalcogenides can be found in the open literature [5,6].

Concerning sensors-based chalcogenides, one of the most common techniques is the usage of chalcogenide glass. The first study was pioneered by [7] on the use of chalcogenide glass-based sensors.

Since then, these glasses have been investigated as potentiometric membranes for the detection of metal cations such as Fe^{3+}, Cd^{2+}, and Pb^{2+}. The ability to synthesize chalcogenide glasses with continuously variable compositions provides a broad range of material properties. Due to the fact that chalcogenide glasses are capable of forming glass with a variety of elements, it is possible to alter the glass compositions to alter their physical, chemical, and electrochemical properties.

In chemistry, chalcogenide glass is a material that contains one or more chalcogens such as sulfur, selenium, and tellurium, but excluding oxygen. As covalently bonded materials, these glasses can be classified as covalent network solids. Since 1976, sensors made of chalcogenide glass have been used to determine the concentrations of heavy metal cations in aqueous media [6]. The primary objective in developing these sensors has always been to achieve the highest possible level of selectivity. The distinguishing characteristics of each of these materials aided in their effective analytical application in industrial research and pollution control. Chalcogenide glasses have been proven as membrane materials with great potential for the detection of heavy metal ions and toxic anions [8–11].

A plethora of research has been carried out related to chemical sensors based on chalcogenide glasses from fundamentals to applications. A quantitative analysis of solutions containing seven ionic species was performed using an array of chalcogenide glass electrodes (heavy metal cations and inorganic anions) in Ref. [12]. The results demonstrated the capability of a sensor array based on chalcogenide glass electrodes combined with artificial neural network data processing to perform multi-component analysis in liquid media. The multi-sensor approach enables the determination of the concentrations of all components in the solution with varying degrees of accuracy. The low selectivity of the single sensors for cadmium and the absence of sensors with a reasonable response for zinc and sulfate did not preclude the analysis of these species as well. Additionally, the sensor's stability was sufficient to estimate concentrations with a reasonable degree of accuracy. Thus, chalcogenide glass multi-sensor arrays can be successfully used for heavy metal cation and inorganic anion sensing in diluted aqueous solutions of complex composition when the non-selective responses are advanced and analyzed using an artificial neural network.

On the other hand, the use of chalcogenide glass chemical sensors (electrodes) in a flow injection setup for the determination of Pb(II), Cr (VI), Cu(II), and Cd(II) was established in Ref. [13]. In their study, flow injection in conjunction with multiple chalcogenide chemical sensors (electrodes) for the quantitative assessment of heavy metals of analytes in solution and the signals were analyzed using a multivariate analysis approach that includes artificial neural networks (Fig. 6.2).

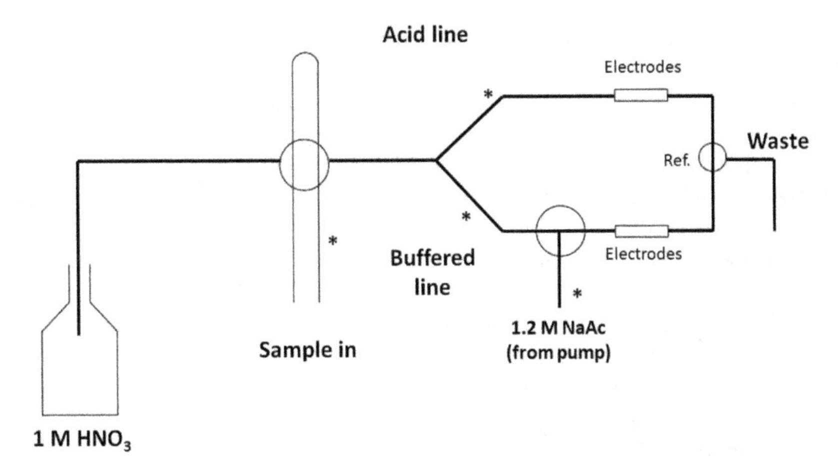

FIGURE 6.2 Flow injection set-up for measuring with multiple chalcogenide glass electrodes (sensor array) [13].

From their work, it can be concluded that the flow injection in conjunction with multi-sensor systems can be used in the same way as other multi-sensor systems to determine analytes. The advantages of flow injection include the ease with which the measurement can be automated and the low load on the chemical sensors during the monitoring process. The multi-sensor flow injection system can be used to monitor industrial emissions of heavy metals (smoke) in real-time.

The fundamental properties of chalcogenide glass of $Ag_x(Ge_{0.25}Se_{0.75})$ 100_x were investigated as a membrane material in ion-selective electrodes in Ref. [14]. The purpose of their work was to investigate how AgGeSe membranes respond to Ag^+, Cu^{2+}, and Fe^{3+} ions. The objective of the work was to investigate how AgGeSe membranes respond to Ag^+, Cu^{2+}, and Fe^{3+} ions. The integration of these elements into the composition of the membrane was studied to determine the effect of metal addition on the selectivity of this mechanism towards Cu and Fe. The fundamental analytic properties of AgGeSe, (AgCu)GeSe, and (AgFe)GeSe systems were discussed. It was found that the chalcogenide glasses that did not contain Ag exhibited Nernstian responses whereas Ag-containing glasses exhibit super-Nernstian. Above a threshold concentration, chalcogenide glasses doped with Ag shift from semiconductor to fast ionic conductor and improve the conductivity to 10^{-5}–10^{-3} S/cm. Furthermore, due to their exceptional glass forming capacity in both bulk and thin films, these materials can be used to build sensitive membranes for micro-sensors in accordance with microelectronic technology. The study also suggested that Ag in the membranes plays a significant role in the sensing mechanism, as indicated by a significant increase in the slope of the electrode calibration function when compared to Nernstian values.

The application of chalcogenide glass chemical sensors is commonly known as environmental sensors to detect heavy metals in water. [15] conducted analytical applications of chemical sensors based on chalcogenide glasses in natural and waste waters from various factories, both in the lab and in situ to trace heavy metals such as copper (II), iron (III), chromium (VI), lead, cadmium, and mercury concentrations. The findings of research using chalcogenide glass chemical sensors agree well with those obtained using atomic absorption spectroscopy and other commonly used methods. In another study by [16], the cross-sensitivity of chalcogenide glass sensors in solutions of heavy metal ions Cu^{2+} of Pb^{2+} and Cd^{2+} at concentrations representative of industrial waste was investigated. The focus of the study was on non-specific sensor response characteristics, particularly cross-sensitivity. The relationship between the sensor material's cation sensitivity and selectivity and the nature of the glass-forming agent, particularly chalcogen, sulfur, or selenium, was observed. The solid-state chalcogenide glass sensors were integrated into an array alongside some conventional It was discovered that this approach enables the determination of all species present in mixed solutions despite the sensors' insufficient selectivity known. Additionally, it appeared possible to determine zinc and sulfate in the absence of selective sensors in the array. The findings of these works have contributed to overcoming the primary difficulty of sensor-based chalcogenide glass sensitivity, which can result in incorrect measurement interpretations. One solution to this problem is to build an electronic tongue-sensing system. The electronic tongue system consists of a collection of several chemical sensors that are processed using advanced mathematical algorithms. It was created to replicate the human gustatory system.

Ion exchange is seen as a very promising technology for removing heavy metals due to its ease of service, economic factors, and high performance [2]. Various chalcogenides based on groups III–V and IV–VI have shown interesting structural and ion-exchange properties in comparison to conventional ion-exchange adsorbents [5], but their preparation and potential use as adsorbents are severely limited due to their vulnerability to acidic conditions. Motivated by a desire to incorporate organic agents into inorganic chalcogenides and produce a series of organic-inorganic hybrid chalcogenides materials with intriguing structures and semiconducting properties, researchers developed a series of organic-inorganic hybrid chalcogenides materials, the ability of intercalation chalcogenide $[CH_3NH_3]_{2x}Mn_xSn_{3-x}S_6 \cdot 0.5H_2O$ ($x = 0.5-1.1$) (CMS) to extract Cd^{2+} and Pb^{2+} from wastewater was studied in depth in Ref. [17]. The aim of the study was that the findings would greatly aid the use of hybrid chalcogenides as adsorbents in wastewater treatment. In this study, the organic cation $CH_3NH_3^+$ was used as a template to replace the inorganic cation K^+ in KMS 1 and synthesized the organic-inorganic hybrid layered materials

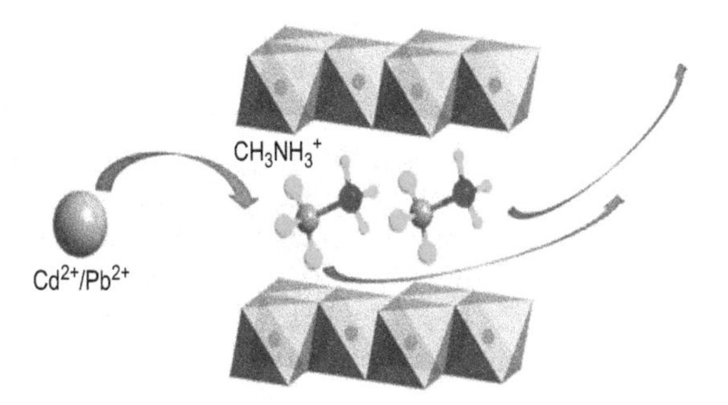

FIGURE 6.3 Synthesized organic-inorganic hybrid layered materials of hybrid chalcogenides [17].

$[CH_3NH_3]_{2x}Mn_xSn_{3-x}S_6 \cdot 0.5H_2O$ ($x = 0.5-1.1$) (CMS) (Fig. 6.3). Hybrid is term as a product created by combining two or more elements. Sn, Mn, S, and methylamine (CH_5N) were used to make $[CH_3NH_3]_{2x}Mn_xSn_{3-x}S_6 \cdot 0.5H_2O$ ($x = 1.0-1.1$) hydrothermally. The CMS was found to be an effective adsorbent for the removal of Cd^{2+} and Pb2 + from water. The CMS exhibited excellent adsorption efficiency for heavy metal ions, including a high capacity for adsorption and a rapid adsorption rate. Adsorption of Cd^{2+}/Pb^{2+} had a negligible effect on pH. The pseudo-second-order kinetic model and the Freundlich equation adequately represented the adsorption mechanism. Additionally, it exhibits high selectivity when used in the presence of competitive ions (Na^+, Ca^{2+}, and Mg^{2+}). Additionally, Cd^{2+}/Pb^{2+} -CMS-based permanent materials can be considered stable. The CMS performs well over a broad pH spectrum, but it was difficult to regenerate the CMS due to the strong affinity between sulfur and heavy metals. The CMS decomposed when a very heavy acidic environment was used. However, the adsorbed materials can be regarded as an excellent permanent waste medium that does not need secondary treatment and does not pose a risk of a heavy metal lease.

Due to their anisotropic property, unique chemical stability, electronic properties, and remarkable structure the scientifically relevant metal chalcogenides (S, Se, and Te) have gotten a lot of attention in the last two decades. Recently, heavy metal removal from water has been discussed in Ref. [18] from the perspective of chalcogenides-based nanomaterials and coupling with metal or non-metal ions. The classification of chalcogenides and their electronic structures, chalcogenides' limitations in photocorrosion as well as their mechanism of action in photocatalytic reduction of heavy metals are presented comprehensively. When coupled with metals or non-metal ions, some chalcogenides (such as ZnS, CuS, CuSe, CdS, RuTe, MoS_2, $MoSe_2$, SnS_2, Ag_2S, $WS_{3,}$) have shown promising coupling results. Extensive research is needed, as well as new ideas and or design strategies in this area gearing towards clean technology.

Heavy metals present in industrial atmospheric particles are released from metallurgical plants such as lead, cadmium, and arsenic. Therefore, continuous monitoring is necessary to reduce the environmental risk and potential effects on the surrounding community. Apart from the application of the chalcogenide glass chemical sensors in the detection of heavy metals in water samples, the chalcogenide glass chemical sensor for cadmium detection in an industrial atmospheric environment was conducted in Ref. [19]. For continuous or quasi-continuous in situ monitoring of industrial atmospheric particles containing heavy metals, advanced chemical sensors based on chalcogenide glasses were developed. The findings for cadmium sensors were presented in the study. The sensors' analytical output was discussed, as well as a laboratory prototype of the monitoring system. It was discovered in the presence of alkali, alkali-earth, and most heavy-metal cations, the sensors tend to be extremely selective.

Quite recently a study was carried out in Ref. [20] to create corresponding planar electrodes in thick film technology in addition to preparing conventional rod electrodes to be used in environmental analysis. The study was motivated to close the gap concerning the conventional rod electrode type of chalcogenide glasses measurement. The enormous demand and technological potential of thick film technology can be put to good use in the production of electrochemical sensors which can be an alternative to the existing technology. In the study, a submicron powder was made from chalcogenides glasses-bulk material to make a sinterable thick film paste. The fundamental tests of the sintering behavior of the chalcogenide glasses under inert conditions were performed. It was found that although the lead selective glasses' material properties prevented the development of processable thick film-pastes, corresponding screen-printed copper selective electrodes were possible. Their measuring capabilities were comparable to those previously recorded for rod-shaped electrodes. The electrode properties were enhanced by optimizing the glass composition.

On the other hand, the recent existing reports of electrochemical sensors based on graphene-metal chalcogenides for the detection of toxic chemicals and heavy metals are summarized in Table 6.2 based on the input from Ref. [21]. Due to various unique physical and chemical properties, transition metal chalcogenides such as sulfides and selenides, are a strong contender for the integration with a conductive graphene host. This has resulted in major improvements in the system's electrochemical efficiency.

Due to the outstanding electrochemical stability and the ability to adjust their sensing characteristics, including selectivity, by changing the glass composition, chalcogenide glass materials are considered one of the most promising sensor membranes in the detection of heavy

TABLE 6.2 Existing studies of electrochemical sensors based on graphene-metal chalcogenides for the detection of toxic chemicals and heavy metals.

References	Electrode material	Analyte	Linear range (μM)
[22]	Au NPs@ MoS_2- rGO/GCE	Hydroquinone Catechol resorcinol	0.1–950 3–560 40–960
[23]	CdS/RGO	CO_2	200–1000 (ppm)
[24]	CuS/RGO	H_2O_2	5–1500
[25]	CdS/RGO	Cu^{2+}	0.5–120
[26]	Cd_xZn 1-x S/ rGO	Cu^{2+}	0.02–20
[27]	CuS I rGO	Hydrazine	1–1000
[28]	CuS-frGO	Cr (IV)	0–200
[29]	Gr/MoS_2/MWCNT	H_2O_2	5–145
[30]	MoS_2/GO/myoglobin hybrid	NO_2	1–3600
[31]	rGO-MoS_2/GCE	NO	0.2–4800
[32]	MoS_2/Gr	Methyl parathion	0.01–1900
[33]	rGO-MoS_2-PEDOT	NO	1–1000
[34]	CuS/GO/MWCNTs	H_2O_2	450–6000
[35]	rGO-MoS_2	NO_2	1–104 ppm

Adapted from A. Pandikumar, P. Rameshkumar, Graphene-Based Electrochemical Sensors for Toxic Chemicals Edited by. 2020.

metals in environmental samples. In addition, with the sensors-based chalcogenides, measurements can be made quickly and easily at the sampling location with just a little effort and without the use of laboratory instruments during the on-site study. As a result, quantitative and semi-quantitative analyses about heavy metal content may be made, for example, in water, atmospheric particles soil, and solid waste.

6.3 Pesticides

Pests have been a big problem for crop production since the dawn of agriculture around 10,000 years ago, and they continue to be so today. Pests are controlled using a variety of methods, including cultural methods such as crop rotation, tillage, and field sanitation, as well as chemical-based pesticides. A pesticide is any material that is used to destroy, repel or control pests in plants or animals. Pesticides are categorized according to many factors, including toxicity (harmful effects), pest organisms killed, and chemical compositions [36]. Table 6.3 summarizes pesticide classifications by adopting information from Ref. [37].

Pest eradication in agriculture and horticulture production is challenging and a major concern for countries. There has been an increase in the development of pesticides that target a wide variety of pests over the last century. The increased volume and frequency of pesticide applications have posed a significant challenge to the targeted pests, causing them to disperse to new environments or adapt to novel conditions. The pest's adaptation to its new environment can be attributed to a variety of mechanisms, including gene mutation, altered population growth rates, and an increase in the number of generations. This has resulted in an increase in pest resurgence and the emergence of pest species resistant to pesticides.

On the hand, environmental exposure and ecotoxicological consequences are used to determine the ecological impact of pesticides. Analysis of pesticides in environmental samples can be done through various methods and environmental sensors. Mass spectrometry is recognized as a highly sensitive and precise technique appropriate for use in environmental organic analysis and is frequently used for pesticide determination. However, this method is quite complex and costly which has motivated the increase in research, experimentation, and technology transfer in this domain to search for alternative techniques over the last few years towards decision support tools for pest management. Electrochemical sensors have received a lot of attention recently as an improvement in the sensors used to detect toxic chemicals including pesticides. The development of graphene-based nanocomposites for electrochemical sensors has paved the way for more sensitive and efficient detection of these chemicals. Due to their unique physical and chemical properties, graphene/metal chalcogenide-based nanocomposites have gained worldwide prominence in recent decades and are being researched for use in pesticide detection and pest management.

6.3.1 Sensors-based chalcogenides for pesticides detection

Concerning pesticide and pest control on crops, improving spraying quality and deposition of pest control products is important. Poor spraying performance results in financial losses, environmental contamination, and a reduction in the biological effectiveness of the applied pesticide. The need for innovative, user-friendly, and readily available decision support tools for measuring spray deposition in field conditions is becoming the main priority. The environmental risks associated

TABLE 6.3 Pesticide classifications.

Pesticide classification	Description
Toxicity	Pesticide toxicity is primarily determined by two factors: dose and duration of exposure. Thus, the amount of the substance involved which is defined as the dose, and the frequency with which the substance is exposed is defined as time. These are considered two distinct types of toxicity which are acute and chronic toxicity.
Pest organisms they kill and pesticide function	*Acaricides*—Substances used to destroy mites and ticks or to prevent them from growing or developing. *Algicide*—Algae-killing or -inhibiting substances. *Fungicides*—Chemicals used to avoid, treat, and remove fungi. *Herbicides*—Chemicals used to kill weeds and other unwanted vegetation *Larvicides*—Inhibit larval growth. *Bactericides*—Compounds isolated from or created by a microorganism, as well as a related chemical synthesized artificially, that are used to destroy or inhibit bacteria in plants or soil. *Insecticides*—are used to control a wide range of insects. *Plant growth regulators*—Substances alter the rate at which plants grow, flower, and reproduce. *Rodenticides*—Chemical substances used to kill rats and other rodents.
Mode of entry	*Systematic*—2,4-Dichlorophenoxyacetic acid (2,4-D), glyphosate. *Non-systemic*—Paraquat, diquat, dibromide. *Stomach poisoning*—Malathion. *Fumigants*—Propylene oxide, dibromo chloropropane. *Repellents*—Citronella, geranium
Chemical Composition	*Insecticides*—Carbamates (Carbaryl), Organochlorine (Endosulfan), Organophosphorus (Monocrotophos), Pyrethroids (permethrin), Neonicotinoids (Imidacloprid), miscellaneous pesticides such as Spinosyns (Spinosad), Benzolureas (diflubenzuron), Antibiotics (abamectin). *Fungicides*—Divided into aliphatic nitrogen fungicides (dodine), amide fungicides (carpropamid), aromatic fungicides (chlorothalonil), dicarboximide fungicides (famoxadone), and dinitrophenol fungicides (dinocap). *Herbicides*—Anilide herbicides (flufenacet), phenoxyacetic herbicides (2,4-D), quaternary ammonium herbicides (Paraquat), chlorotriazine herbicides (atrazine), sulfonylurea herbicides (chlorimuron). *Rodenticides*—Inorganic rodenticides (zinc phosphide, aluminum phosphide) and coumarin rodenticides (organic) are the two types of rodenticides (bromadiolone, coumatetralyl).

Adapted from M.M. Akashe, U.V Pawade, A.V Nikam, Classification of pesticides: a review, Int. J. Res. Ayurveda Pharm., 9, 4, pp. 144–150, 2018.

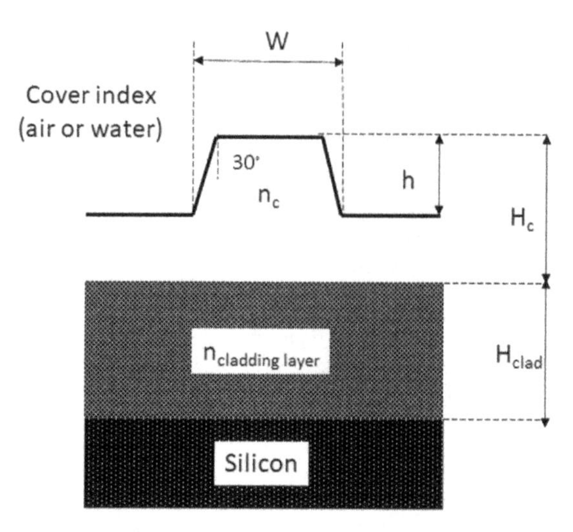

FIGURE 6.4 Waveguide transverse structure [43].

with agricultural operations could be greatly reduced if these needs are met. Numerous methods have been developed to quantify spray deposition in field conditions for crop pest eradication in agriculture and horticulture [38]. These include the usage of water-sensitive paper cards [39], spectroscopy [40], and colorimetry [41]. These methods, on the other hand, are both expensive and time-consuming. The need for quick, automated devices is essential. This has prompted the development of electronic, resistance-based sensor array and data acquisition in environmental monitoring for pest control [42] including optical planar waveguide chalcogenide sensors. A study by [43] was performed concerning chalcogenide rib waveguides for the characterization of spray deposits. In their study, the focus was placed on the feasibility of using optical sensors to track the quantity and distribution of liquid spray. To do so, rib waveguides operating at 1.55 m were designed and developed, with a water absorption coefficient of about 50 cm (Fig. 6.4). They were made up of two $Ge_{25}Se_{65}Te_{10}$ and $Ge_{25}Se_{55}Te_{20}$ chalcogenide layers. The effect of water droplet deposition on transmitted light intensity was investigated both experimentally and numerically. From the study, it was discovered that the direction of the drop on the waveguide did not affect the strength measured at the guide outlet, according to both experimental and simulation results. On the other hand, they emphasized the connection between light intensity at the outlet and the volume and number of deposited droplets: the larger the drop volume and the greater the number of drops, the lower the intensity. The outcome of the study paves the way for the creation of chalcogenide sensors to detect spray deposits, specifically pesticide depositions in agriculture.

In another study, [44] developed low-cost micro-sensors that can be used in the field to quickly measure the plant's leaf coverage surface, which is often used as a key measurement of the spraying process to optimize spray quality. In their work, the light transmission of rib optical waveguides made of Ge-Se-Te chalcogenide films was studied in response to the deposition of demineralized water droplets on their surface. The transmission spectra at the output of the waveguides were recorded before (reference) and after drop deposition using a dedicated spectrophotometric bench in the wavelength range between 1200 and 2000 nm. The presence of a hollow at 1450 nm has been observed in the relative transmission spectra. The first overtone of the O–H stretching vibration in water refers to this. This finding suggests that the optical intensity decrease observed after droplet deposition is due in part to the light energy carried by the directed mode evanescent field being absorbed by water. As a result, the probe based on Ge-Se-Te rib optical waveguides is sensitive across the entire volume range tested which ranged from 0.1 to 2.5 L. As a result of multivariate methods, the statistics of the measurements and the predictive character of the transmission spectra could be analyzed. It verified the measurement system's sensitivity to water absorption, and the predictive model allowed for droplet volume prediction on an independent collection of measurements, with a correlation of 66.5% and precision of 0.39 L. Further research will be needed to determine the best optical waveguide structure to create, low-cost and compact laser sources and spectrometers, as well as optimized packing and connectors, before the technology can be used to develop a sensor for use in real-world conditions.

On the other hand, land deposits during the application of pesticides to crops are one of the mechanisms by which the environment can be contaminated. The retention rate of sprayed droplets is critical for both phytosanitary treatment effectiveness and the number of pesticides lost on the field which could be considered for future studies in this field. [45] reported on a study of measurement techniques and simulation studies in the characterization of spray droplet patterns to meet the target and optimize spray crop protection performance. The study provided an overview of the factors that influence spray droplet activity, the processes involved in sprayed drop fate, and the most common methods for determining pesticide deposits on the ground and plant retention. Thus, the development of chalcogenide sensors for the detection of spray deposits is also a possibility in this domain.

The development of mid-infrared sensors for biochemical molecule detection is a significant challenge. Chalcogenide glass sensors have been successfully implemented for the detection of biochemical molecules in various fields of applications including water pollution, microbiology, and medicine [46]. The absorption bands related to the vibrations of organic molecules are found in the mid-infrared range (4000–400 cm^{-1}) including pesticides. Concerning this, chalcogenide glasses are primarily studied and used for their high refractive index and wide infrared clarity which is up to 15 m for selenide glasses. [47] conducted a study in which they synthesized and characterized chalcogenide thin films to create mid-IR optical waveguides. Two $(GeSe_2)100-x(Sb2Se3)x$ chalcogenide glasses, where $x = 10$ and 50, were chosen for their strong mid-IR clarity, high stability against crystallization, and refractive index contrast suitable for mid-IR waveguiding and were prepared using the traditional melting and quenching method and then used for RF magnetron sputtering deposition in their analysis. The study's findings were positive in terms of developing an integrated optical sensor that operates in the mid-infrared and could be used to detect pesticides.

6.4 Extreme environmental conditions

Various detection materials have been thoroughly studied for decades in the hopes of discovering a stronger substance or mechanism for extreme environment sensing such as high temperature and radiation. The optical fiber sensor has been proposed as a viable alternative to conventional real-time monitoring for monitoring various parameters such as temperature, pressure, strain, and condition of coal-fired power plants operating in extreme environments. A recent study by [48] presented a novel, accurate, real-time temperature monitoring device based on the phase change of chalcogenide glasses at specific, well-defined temperatures. The developed sensor is shown in Fig. 6.5. The sensor architecture took advantage of the heat-induced phase change (amorphous to crystalline) property of chalcogenide glasses, which altered the material's optical properties rapidly (80–100 ns). From the study, it was discovered that the sensor promised a highly reliable, compact, and lightweight sensor network that is ideal for use in high-temperature and

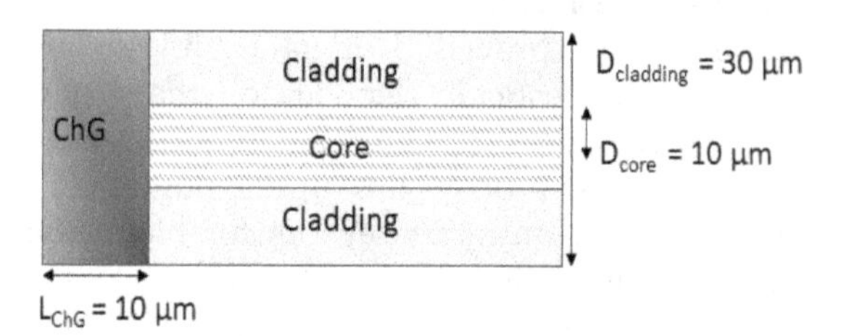

FIGURE 6.5 Schematic cross-section of proposed ChG-capped optical tip-based temperature sensor [48].

TABLE 6.4 Studied composition of synthesized glasses in amorphous and crystalline phases [48].

	Refractive index		Temperature		
Composition	**Amorphous**	**Crystalline**	T_g	T_o	T_c
$Ge_{30}S_{70}$	2.17406 + i0	1.77269 + i0.11865	402	572	
$Ge_{40}S_{60}$	2.67680 + i0	2.72309 + i0.17664	355	408	413
				480	489
$Ge_{33}S_{67}$	2.31779 + i8.28 × 10 − 6	1.92455 + i0.02458	435	644	694
$Ge_{30}Se_{70}$	2.37646 + i4.06 × 10 − 5	3.12455 + i0.25837	334.8	440.9	470.4
$Ge_{40}Se_{60}$	2.63104 + i0.00575	3.10991 + i0.211	343.7	446.6	472.3
$Ge_{33}Se_{67}$	2.38753 + i0.00402	2.30756 + i0.02011	396.3	485.4	527.7

high-radiation environments. Significant compositional and structural changes in Ge-rich compositions following crystallization resulted in more expressed device effects, which is critical when considering material choices for building devices. The proposed device was reusable after an electrical field was applied at room temperature to the chalcogenide glass-coated fiber tip, causing Joule heating and melting of the chalcogenide glasses. Due to rapid cooling from the fiber core, which had a much larger volume than the chalcogenide glass film, the melt solidifies in an amorphous state.

The simulation models incorporate the measured complex refractive indices of in-house synthesized Ge-Se and Ge-S compositions in amorphous and crystalline phases. The T_g, T_o, and T_c values for all synthesized and studied compositions are summarized in Table 6.4. Among the large ChG family, the Ge-Se and Ge-S systems were chosen for the proposed device because they are thermally stable between 400°C and 650°C and their glass-forming regions are quite extensive, providing a substantial number of compositions to work with.

In a radiation sense, it has been an increasing demand for a smaller and more reliable material or system that can perform the same functions as bulkier Geiger counters and other measurement options. These smaller sensors are developed to fulfill the requirement for easy, inexpensive, and accurate radiation dose measurement. The use of thin film chalcogenide glass, which has unique properties such as high thermal stability and high sensitivity to short wavelength radiation, becomes essential. Concerning this, [49] developed a low-cost, high-performance microelectronic device that reacts to gamma radiation to produce an easily measured change in electrical resistance. This radiation sensor

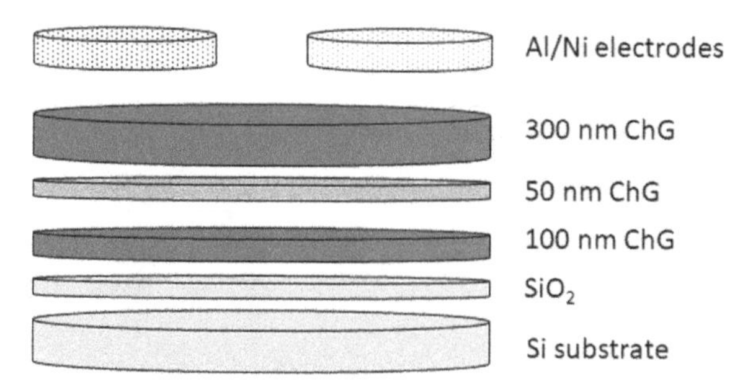

FIGURE 6.6 Cross section of the chalcogenide glass [49].

was a two-terminal microdevice with an active region consisting of a chalcogenide glass which was fabricated based on amorphous films (Fig. 6.6). The sensor was fabricated on non-planar and versatile substrates, extending its range of operation and allowing for calibration and reuse.

The key issues with polymer membranes, however, are their durability and short life span. Chalcogenide glass can be a safer alternative to polymer membranes for continuous work in extreme environments, due to its durability and flexibility. The current state of research on chalcogenide glass calls for lowering both limits of detection and cross-sensitivity, which could be achieved by enhancing glass synthesis with new glass compositions while preserving the thin film membrane's stoichiometry.

6.5 Summary

There has been tremendous work on the investigation of chalcogenide glass sensors have been developed since the 1970s for numerous applications because of their chemical stability, ease of use, and data processing, Current research on sensor-based chalcogenide glasses in the open literature covers a wide range of environmental applications, including the identification of heavy metals and pesticides in different environmental segments such as water, wastewater, and the atmosphere. These sensors have the potential to be low-cost, high-performance, and capable of producing an easily measured change in resistance. Since the chalcogenide glass family contains a wide range of materials, there are more options for tailoring the sensor's sensitivity to specific environmental applications. Because of the various and potentially comprehensive applications, the authors conclude that chalcogenide glass sensors will rise in popularity for environmental applications. The availability of high-quality, low-loss, high-strength single-mode and multimode materials would certainly enhance existing technologies while also enabling new ones. Furthermore, the glasses have the potential to be used to fabricate solid-contact sensors. They exhibit a high degree of chemical stability, resulting in chemical durability. It is possible to convert glasses into thin films to fabricate miniature sensors Additionally, the low impedance of a thin film in comparison to the bulk material allows for the use of poor conducting glass. In a nutshell, the future of chalcogenide glasses sensors appears to be very promising.

References

[1] P.B. Tchounwou, C.G. Yedjou, A.K. Patlolla, D.J. Sutton, Heavy metal toxicity and the environment, Mol. Clin. Environ. Toxicol. (2012) 133–164. Available from: https://doi.org/10.1007/978-3-7643-8340-4_6.

[2] N. Fourati, N. Blel, Y. Lattach, N. Ktari, M.E. Zerrouki, Chemical and Biological Sensors from Conducting and Semiconducting Polymers, Elsevier, 2016. Available from: https://doi.org/10.1016/B978-0-12-803581-8.01733-1.

[3] A.G.-M. Ferrari, P. Carrington, S.J. Rowley-Neale, C.E. Banks, Recent advances in portable heavy metal electrochemical sensing platforms, Environ. Sci. Water Res. Technol. 6 (10) (2020) 2676–2690. Available from: https://doi.org/10.1039/d0ew00407crsc.li/es-water.

[4] R.K. Soodan, Y.B. Pakade, A. Nagpal, J.K. Katnoria, Analytical techniques for estimation of heavy metals in soil ecosystem: a tabulated review, Talanta 125 (2014) 405–410. Available from: https://doi.org/10.1016/j.talanta.2014.02.033.

[5] T.V. Moreno, et al., Potentiometric sensors with chalcogenide glasses as sensitive membranes: a short review, J. Non. Cryst. Solids 495 (2018) 8–18. Available from: https://doi.org/10.1016/j.jnoncrysol.2018.04.057.

[6] M.J. Schöning, J.P. Kloock, About 20 years of silicon-based thin-film sensors with chalcogenide glass materials for heavy metal analysis: technological aspects of fabrication and miniaturization, Electroanal. An. Int. J. Devoted Fundam. Pract. Asp. Electroanal. 19 (19-20) (2007) 2029–2038. Available from: https://doi.org/10.1002/elan.200703955.

[7] C.T. Baker, I. Trachtenberg, Ion selective electrochemical sensors—Fe^{3+}, Cu^{2+}, J. Electrochem. Soc. 118 (4) (1971) 571.

[8] N. Tohge, M. Tanaka, Chalcogenide glass electrodes sensitive to heavy metal ions, J. Non. Cryst. Solids 80 (1) (1986) 550–556. Available from: https://doi.org/10.1016/0022-3093(86)90445-X.

[9] Y.G. Vlasov, E.A. Bychkov, Ion-selective chalcogenide glass electrodes, Ion-Selective Electrode Rev. 9 (1) (1987) 5–93.

[10] E. Bychkov, Spectroscopic studies of chalcogenide glass membranes of chemical sensors: local structure and ionic response, Sens. Actuators B Chem. 27 (1–3) (1995) 351–359. Available from: https://doi.org/10.1016/0925-4005(94)01616-P.

[11] M. Miloshova, et al., Tracer and surface spectroscopy studies of sensitivity mechanism of mercury ion chalcogenide glass sensors, Sens. Actuators B Chem. 57 (1–3) (1999) 171–178. Available from: https://doi.org/10.1016/S0925-4005(99)00078-7.

[12] C. Di Natale, et al., Multicomponent analysis of heavy metal cations and inorganic anions in liquids by a non-selective chalcogenide glass sensor array, Sens. Actuators B Chem. 34 (1) (1996) 539–542. Available from: https://doi.org/10.1016/S0925-4005(96)01925-9.

[13] J. Mortensen, A. Legin, A. Ipatov, A. Rudnitskaya, Y. Vlasov, K. Hjuler, A flow injection system based on chalcogenide glass sensors for the determination of heavy metals, Anal. Chim. Acta 403 (1) (2000) 273–277. Available from: https://doi.org/10.1016/S0003-2670(99)00544-9.

[14] J.M. Conde Garrido, F. Macoretta, M.A. Ureña, B. Arcondo, Application of Ag–Ge–Se based chalcogenide glasses on ion-selective electrodes, J. Non. Cryst. Solids 355 (37) (2009) 2079–2082. Available from: https://doi.org/10.1016/j.jnoncrysol.2008.12.026.

[15] A.V. Legin, E.A. Bychkov, Y.G. Vlasov, Analytical applications of chalcogenide glass chemical sensors in environmental monitoring and process control, Sens. Actuators B Chem. 24 (1) (1995) 309–311. Available from: https://doi.org/10.1016/0925-4005(95)85067-8.

[16] A.V. Legin, Y.G. Vlasov, A.M. Rudnitskaya, E.A. Bychkov, Cross-sensitivity of chalcogenide glass sensors in solutions of heavy metal ions, Sens. Actuators B Chem. 34 (1) (1996) 456–461. Available from: https://doi.org/10.1016/S0925-4005(96)01852-7.

[17] J.-R. Li, X. Wang, B. Yuan, M.-L. Fu, H.-J. Cui, Robust removal of heavy metals from water by intercalation chalcogenide $[CH_3NH_3]_{2x}Mn_xSn_{3-x}S_6 \cdot 0.5H2O$, Appl. Surf. Sci. 320 (2014) 112–119. Available from: https://doi.org/10.1016/j.apsusc.2014.09.057.

[18] M. Madkour, H.A. El Nazer, Y.K. Abdel-Monem, 11 - Use of chalcogenides-based nanomaterials for photocatalytic heavy metal reduction and ions removal, in: P. Khan (Ed.), Micro and Nano Technologies, Elsevier, 2021, pp. 261–283. Available from: https://doi.org/10.1016/B978-0-12-820498-6.00011-1.

[19] M. Milochova, M. Kassem, E. Bychkov, Chalcogenide glass chemical sensor for cadmium detection in industrial environment, ECS Trans. 50 (12) (2013) 357.

[20] U. Enseleit, M. Berthold, C. Feller, U. Partsch, S. Körner, W. Vonau, Chalcogenide glass based heavy metal sensors, Sens. Transducers 219 (1) (2018) 1–8.

[21] S. Stanly John Xavier, T.S.T. Balamurugan, S. Ramalingam, R. Ramachandran, N.S.K. Gowthaman, Graphene-carbon nitride based electrochemical sensors for toxic chemicals, Mater. Res. Foundations 82 (2020), 243–275.

[22] G. Ma, H. Xu, M. Wu, L. Wang, J. Wu, F. Xu, A hybrid composed of MoS_2, reduced graphene oxide and gold nanoparticles for voltammetric determination of hydroquinone, catechol, and resorcinol, Microchim. Acta 186 (11) (2019) 1–9. Available from: https://doi.org/10.1007/s00604-019-3771-4.

[23] A. Hasani, H. Sharifi Dehsari, A. Amiri Zarandi, A. Salehi, F.A. Taromi, H. Kazeroni, Visible light-assisted photoreduction of graphene oxide using CdS nanoparticles and gas sensing properties, J. Nanomater. 2015 (2015). Available from: https://doi.org/10.1155/2015/930306.

[24] J. Bai, X. Jiang, A facile one-pot synthesis of copper sulfide-decorated reduced graphene oxide composites for enhanced detecting of H_2O_2 in biological environments, Anal. Chem. 85 (17) (2013) 8095–8101. Available from: https://doi.org/10.1021/ac400659u10.1021/ac400659u.

[25] I. Ibrahim, H.N. Lim, N.M. Huang, A. Pandikumar, Cadmium sulphide-reduced graphene oxide-modified photoelectrode-based photoelectrochemical sensing platform for copper (II) ions, PLoS One 11 (5) (2016) e0154557. Available from: https://doi.org/10.1371/journal.pone.0154557.

[26] J. Yan, et al., One-pot synthesis of $Cd_xZn_{1-x}S$–reduced graphene oxide nanocomposites with improved photoelectrochemical performance for selective determination of Cu^{2+}, RSC Adv. 3 (34) (2013) 14451–14457. Available from: https://doi.org/10.1039/C3RA41118D.

[27] Y.J. Yang, W. Li, X. Wu, Copper sulfide | reduced graphene oxide nanocomposite for detection of hydrazine and hydrogen peroxide at low potential in neutral medium, Electrochim. Acta 123 (2014) 260–267. Available from: https://doi.org/10.1016/j.electacta.2014.01.046.

[28] P. Borthakur, M.R. Das, S. Szunerits, R. Boukherroub, CuS decorated functionalized reduced graphene oxide: a dual responsive nanozyme for selective detection and photoreduction of Cr (VI) in an aqueous medium, ACS Sustain. Chem. Eng. 7 (19) (2019) 16131–16143. Available from: https://doi.org/10.1021/acssuschemeng.9b03043.

[29] M. Govindasamy, V. Mani, S.M. Chen, R. Karthik, K. Manibalan, R. Umamaheswari, MoS_2 flowers grown on graphene/carbon nanotubes: a versatile substrate for electrochemical determination of hydrogen peroxide, Int. J. Electrochem. Sci. 11 (4) (2016) 2954–2961. Available from: https://doi.org/10.20964/11040295410.20964/110402954.

[30] J. Yoon, et al., Electrochemical nitric oxide biosensor based on amine-modified MoS_2/graphene oxide/myoglobin hybrid, Colloids Surf. B Biointerfaces 159 (2017) 729–736. Available from: https://doi.org/10.1016/j.colsurfb.2017.08.033.

[31] J. Hu, et al., Synthesis and electrochemical properties of rGO-MoS_2 heterostructures for highly sensitive nitrite detection, Ionics) 24 (2) (2018) 577–587. Available from: https://doi.org/10.1007/s11581-017-2202-y.

[32] M. Govindasamy, S.-M. Chen, V. Mani, M. Akilarasan, S. Kogularasu, B. Subramani, Nanocomposites composed of layered molybdenum disulfide and graphene for highly sensitive amperometric determination of methyl parathion, Microchim. Acta 184 (3) (2017) 725–733. Available from: https://doi.org/10.1007/s00604-016-2062-6.

[33] R. Madhuvilakku, S. Alagar, R. Mariappan, S. Piraman, Glassy carbon electrodes modified with reduced graphene oxide-MoS_2-poly (3, 4-ethylene dioxythiophene) nanocomposites for the non-enzymatic detection of nitrite in water and milk, Anal. Chim. Acta 1093 (2020) 93–105. Available from: https://doi.org/10.1016/j.aca.2019.09.043.

[34] W. Jin, Y. Fu, W. Cai, In situ growth of CuS decorated graphene oxide-multiwalled carbon nanotubes for ultrasensitive H_2O_2 detection in alkaline solution, N. J. Chem. 43 (8) (2019) 3309–3316. Available from: https://doi.org/10.1039/C8NJ06134C.

[35] A. Mukherjee, L.R. Jaidev, K. Chatterjee, A. Misra, Nanoscale heterojunctions of rGO-MoS_2 composites for nitrogen dioxide sensing at room temperature, Nano Express 1 (1) (2020) 10003. Available from: https://doi.org/10.1088/2632-959X/ab7491.

[36] N.A. Ahmad, A. Salehabadi, S.A. Muhammad, M.I. Ahmad, Potential risk and occupational exposure of pesticides among rice farmers of a village located in Northern Peninsular of Malaysia, Expo. Heal. 12 (4) (2020) 735–749. Available from: https://doi.org/10.1007/s12403-019-00333-9.

[37] M.M. Akashe, U.V. Pawade, A.V. Nikam, Classification of pesticides: a review, Int. J. Res. Ayurveda Pharm. 9 (4) (2018) 144–150.

[38] T.G. Crowe, D. Downey, D.K. Giles, Digital device and technique for sensing distribution of spray deposition, Trans. ASAE 48 (6) (2005) 2085–2093.

[39] M. Cunha, C. Carvalho, A.R.S. Marcal, Assessing the ability of image processing software to analyse spray quality on water-sensitive papers used as artificial targets, Biosyst. Eng. 111 (1) (2012) 11–23. Available from: https://doi.org/10.1016/j.biosystemseng.2011.10.002.

[40] R.C. Derksen, R.L. Gray, Deposition and air speed patterns of air-carrier apple orchard sprayers, Trans. Am. Soc. Agric. Eng. 38 (1) (1995) 5–11. Available from: https://doi.org/10.13031/2013.2780510.13031/2013.27805.

[41] W.C. Hoffmann, M. Salyani, Spray deposition on citrus canopies under different meteorological conditions, Trans. ASAE 39 (1) (1996) 17–22. Available from: https://doi.org/10.13031/2013.27475.

[42] M.A. Kesterson, J.D. Luck, M.P. Sama, Development and preliminary evaluation of a spray deposition sensing system for improving pesticide application, Sensors 15 (12) (2015) 31965–31972. Available from: https://doi.org/10.3390/s151229898.

[43] A. Taleb Bendiab, et al., Chalcogenide rib waveguides for the characterization of spray deposits, Opt. Mater. 86 (2018) 298–303. Available from: https://doi.org/10.1016/j.optmat.2018.10.021.

[44] A. Taleb Bendiab, et al., Coupling waveguide-based micro-sensors and spectral multivariate analysis to improve spray deposit characterization in agriculture, Sensors 19 (19) (2019) 4168. Available from: https://doi.org/10.3390/s19194168.

[45] A. Allagui, H. Bahrouni, Y. M'Sadak, Deposition of pesticide to the soil and plant retention during crop spraying: the art state, J. Agric. Sci. 10 (2018) 12. Available from: https://doi.org/10.5539/jas.v10n12p104.

[46] F. Charpentier, et al., CO_2 detection using microstructured chalcogenide fibers, Sens. Lett. 7 (5) (2009) 745–749. Available from: https://doi.org/10.1166/sl.2009.1142.

[47] E. Baudet, et al., Infrared sensor for water pollution and monitoring, Optical Sens. 10231 (2017) 102310S. Available from: https://doi.org/10.1117/12.2264899. 2017.

[48] B. Badamchi, A.-A. Ahmed Simon, M. Mitkova, H. Subbaraman, Chalcogenide glass-capped fiber-optic sensor for real-time temperature monitoring in extreme environments, Sensors 21 (5) (2021) 1616. Available from: https://doi.org/10.3390/s21051616.

[49] M. Mitkova, D. Butt, M. Kozicki, H. Barnaby, Chalcogenide Glass Radiation Sensor; Materials Development, Design and Device Testing, UT-Battelle LLC/ORNL, Oak Ridge, TN (Unted States), 2013. Available from: https://doi.org/10.2172/1082961.

CHAPTER

7

Biological molecule sensors

7.1 DNA

In biology, DNA is an acronym for deoxyribonucleic acid, which is an organic molecule with a complex molecular structure that is present in all prokaryotic and eukaryotic cells, as well as in a variety of viruses. The genetic information encoded by DNA is responsible for the transfer of inherited characteristics. It is found in all living things and conveys genetic information. Known as a double helix (Fig. 7.1), the DNA molecule is composed of two strands that spiral around one another to form the structure of the molecule itself. Nucleotides are the building blocks of DNA. A nucleotide has two components: a backbone, made from the sugar deoxyribose and phosphate groups, and nitrogenous bases, known as cytosine, thymine, adenine, and guanine. The genetic code is produced by rearranging the nucleotides. There are four bases attached to each sugar: adenine (A), cytosine (C), guanine (G), and thymine (T) (T). The two strands are linked together by base pairings; adenine forms a bond with thymine, and cytosine forms a bond with guanine, respectively. The sequence of bases along the backbones of proteins and RNA molecules acts as a set of instructions for building the molecules.

In recent years, sensitive detection of DNA sequence has become more significant in the prevention and treatment of genetic abnormalities, the prognosis of cancer, as well as the treatment of bacterial and viral infections [1]. To enable early and precise illness diagnosis, a great deal of interest has been generated in the development of simple, selective, and sensitive techniques for DNA sequence and mutant gene analysis, which are intended to be simple, selective, and sensitive. Due to its inherent advantages of high sensitivity, compatibility with miniaturization, ease of operation, and cheap cost, electrochemical assays have been the most preferred technique for DNA detection [2]. However, the

Metal Chalcogenide Biosensors
DOI: https://doi.org/10.1016/B978-0-323-85381-1.00005-2

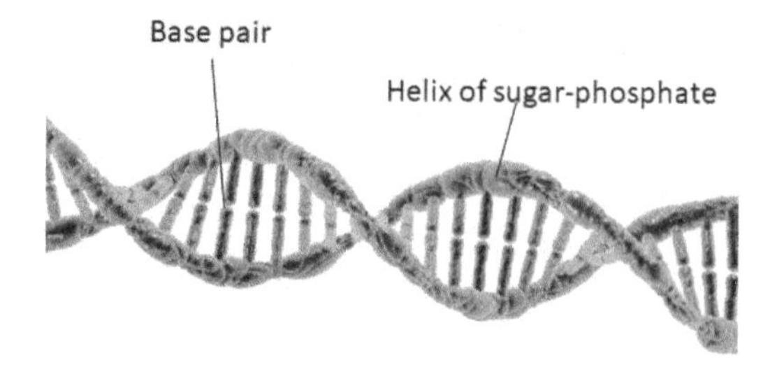

FIGURE 7.1 Deoxyribonucleic acid (DNA).

electrochemical response of the vast majority of bare electrodes is suboptimal which limits the direct detection of DNA sequences in biological substances. Concerning this, 2D transition-metal chalcogenides [such as titanium disulfide (TiS2), tungsten disulfide (WS2), vanadium disulfide (VS2), and molybdenum disulfide (MoS2)], metal oxides [manganese(IV) oxide (MnO2) and nickel(II) hydroxide (-Ni(OH)2] have recently gained increased attention because they typically exhibit unique physical and electrical properties due to their two-dimensional ultrathin atomic layer structure and large surface area, making them extremely interesting for use in nanoelectronics, optoelectronics, and energy harvesting, as well as for use as a promising supporting material in biosensing platform [3].

7.1.1 DNA-linked chalcogenide studies

Nowadays, producing a well-defined low-dimensional nanostructure using deoxyribonucleic acid (DNA)-based nanotechnology is a hot topic of research for many research communities, particularly for bio-hybrid nanomaterials, which combine inorganic materials with biomolecules [4]. Numerous biological components, including DNA, have been extensively used as bio-templates to build particular nanostructures of inorganic-bio hybrid materials [5]. Several researchers testified about DNA-linked chalcogenide quantum dots and their applications [4].

Among the various semiconductor chalcogenide systems, CdS has garnered considerable interest due to its fundamental and effective properties, which include a direct bandgap, excellent chemical and thermal stability, a relatively low work function, excellent transport properties, high electronic mobility, and piezoelectricity [6,7]. As a result, CdS is an enthralling material with well-established applications in light-emitting diodes, photodetectors, waveguides, sensors, photovoltaics, photoelectrochemical cells, and nanogenerators, among others [8,9]. Numerous techniques have been devised for the fabrication of cadmium chalcogenides nanowires [3]. CdS quantum dots have recently been considered a potential material for biological applications, particularly for multiplex labeling and detection of DNAs.

Cadmium sulfide nanoparticles were synthesized in a study by Kulkarni et al. [10] in aqueous and non-aqueous environments, new. DNA was added during the nanoparticles' manufacturing, resulting in cadmium-rich nanoparticles that form a stable combination with DNA. These particles show intense fluorescence, the wavelength of which is determined by the media in which they are manufactured. When CdS nanoparticles interact with proteins, their fluorescence peak intensity increases significantly. Such CdS nanoparticles may be effective as a protein sensor.

Schottky junction devices by using bio-molecule DNA template-based one-dimensional CdS-nanostructures were investigated [4]. The electrical characteristics of biomolecule DNA-template-based one-dimensional nanowires (NWs)-CdS/Au and nanoparticles (NPs)-CdS/Au devices grown on Indium Tin Oxide (ITO) glass substrates were studied [4]. It was found the NWs-CdS/Au device exhibits a significant increase in current flow and a significant decrease in threshold voltage (55 mV) when compared to NPs (190 mV) and previously published bulk-CdS/Au (680 mV) devices. While the non-linear/asymmetric current-voltage ($I-V$) characteristic identifies the CdS/Au junction as a Schottky device, the huge ideality factor of roughly 24 seen in the NWs-CdS/Au device may be a result of the DNA-assembled organic process CdS-semiconductor. The discovery of considerable changes in the electrical characteristics of DNA-assembled NWs-based Schottky junctions may aid in the development of more complex and multispecific biosensors for medical applications. On the other hand, Al-Ta'ii et al. [11] studied the electrical properties of an Au/DNA/ITO device constructed by self-assembly were determined utilizing current-voltage ($I-V$) characteristic measurements at room temperature during an alpha bombardment. The Au/DNA/ITO structure's barrier height, ideality factor, series resistance, and Richardson constant were estimated using the traditional thermionic emission model, Cheung and Cheung's approach, and Norde's methodology. Apart from displaying a strongly rectifying (diode) feature, orderly oscillations in several electrical properties of the Schottky structure were found. Alpha radiation has a significant effect on series resistance, whereas barrier height, ideality factor, and interface state density parameters respond linearly. The barrier height was computed using $I-V$ measurements and was found to be 0.7284 eV for non-radiated samples, increasing to around 0.7883 eV in 0.036 Gy, indicating a rise for all doses. Additionally, we illustrate the influence of hypersensitivity by examining the link between the series resistance for the three approaches, the ideality factor, and low-dose radiation. As a result of the findings, sensitive alpha particle detectors utilizing an

Au/DNA/ITO Schottky junction sensor can be created. Further research utilizing various DNA template sequences may aid in controlling the size, shape, and performance of chalcogenide semiconductor devices. The current work advances the field of self-assembly functional nanoscale electronic devices and nanoelectronics by characterizing DNA-based nanoelectronics.

The nanofabrication of inorganic nanostructures using DNA has the potential to be used in electronics, catalysis, and plasmonics. Although previous DNA metallization techniques have resulted in conductive DNA nanostructures, the usage of semiconductors and the construction of well-connected nanoscale metal-semiconductor junctions on DNA nanostructures are still in their infancy. In this regard, [12] demonstrated the creation of several electrically linked metal-semiconductor junctions on individual DNA origami using the site-specific binding of gold and tellurium nanorods. Au nanorods were attached to DNA origami by DNA hybridization, whereas Te nanorods were attached via electrostatic contact. By filling the spaces between Au and Te nanorods, electroless gold plating was employed to generate nanoscale metal-semiconductor interactions. Electrical characterization at two points demonstrated that the Au-Te-Au junctions were electrically coupled, exhibiting current–voltage characteristics similar to those of a Schottky junction. The nanofabrication of metal-semiconductor junctions using DNA demonstrates the strength of this bottom-up method in nanoelectronics.

Following the discovery of graphene, transition metal dichalcogenides have attracted considerable attention because these materials have the potential for use in a variety of applications. Several studies on two-dimensional gallium selenide (GaSe), a member of the post-transition-metal chalcogenides (PTMCs) family, have been published in the literature [13]. The GaSe monolayers consist of two sub-layers of gallium (Ga) atoms sandwiched between two sub-layers of selenium (Se) atoms, with the gallium (Ga) atoms forming the outermost layer. The exceptional features of GaSe, as well as the fact that it is well suited for bio-sensing applications, have motivated us to examine the possibility of GaSe for next-generation DNA sequencing devices. The examination of critical bimolecular compounds (amino acids and nucleobases) in the human body is extremely significant to extract and sequencing the genetic information contained within them. The interaction of DNA nucleobases with monolayer GaSe was investigated utilizing the vdW functional in a DFT framework [14]. On the GaSe monolayer, they discovered that nucleobases are physisorbed. $C > T > G > A$ is the order of binding energy per atom. The room temperature recovery time for T + GaSe is calculated to be a maximum of 113.88 s, indicating that the GaSe-based devices are reusable. Within the simulated STM experiments, the variation in the electrical structures of GaSe was recorded. It was concluded from their study that, the anisotropic optical response of GaSe monolayers in the UV range is extremely advantageous for creating polarized optical sensors. Their findings reveal that a GaSe monolayer may be used to create reusable DNA sequencing devices for use in biotechnology and medicine.

Due to their outstanding structural and electrical features, two-dimensional (2D), transition-metal chalcogenides have recently gained increased interest. Huang and coworkers [15] devised an electrochemical assay for the identification of particular DNA sequences by building a capture DNA probe on a modified AuNPs/CuS-AB film and employing $[Fe(CN)_6]3/4$ as an indicator. Due to their high conductivity and huge electroactive surface area, CuS-AB composites and AuNPs enabled signal amplification for electrochemical detection. The new test demonstrated great sensitivity and selectivity for the target DNA sequence under appropriate experimental conditions. In other studies, Tan et al. [16] demonstrated the high-yield and scalable manufacture of ultrathin 2D ternary chalcogenide nanosheets in solution, including Ta_2NiS_5 and Ta_2NiSe_5. The resultant Ta_2NiS_5 and Ta_2NiS_5 nanosheets are between tens of nanometers to a few micrometers in size. Notably, the yield of single-layer Ta_2NiS_5 nanosheets is extremely high, around 86%. As a proof-of-concept, the single-layer Ta_2NiS_5 is employed as a new fluorescence sensing platform for the highly selective and sensitive detection of DNA (with a detection limit of 50 pM). These solution-processable, high-yield and large-amount ternary chalcogenide nanosheets may also find use in electrocatalysis, supercapacitors, and electrical devices.

7.2 Protein

Proteins are a group of macromolecules inside cells consisting of polypeptides. Any type of protein is distinguished by a special polypeptide sequence. In most cases, a special protein's presence or concentration value can assist in disease rate recognition. This section studies novel findings on fabricated protein biosensors by some metal chalcogenides such as cadmium and zinc chalcogenides (CdX, ZnX, X = S, Se). Four components necessary for biosensor fabrication are related to each other in Table 7.1, based on several protein biosensors that are discussed in the section.

Xu et al. [17] prepared an efficient photoelectrochemical (PEC) immunosensor for amyloid beta (Aβ (1 ~ 42)) peptides detection. These peptides have a great neurotoxic effect on Alzheimer's disease (AD) progression. Sensitized Bi_2WO_6/CdS electrode to Mn^{2+} doped CdSe (Mn: CdSe) was measured for the protein. The three-dimensional

TABLE 7.1 Four constituent components of the biosensors for protein detection.

	Component 1: Metal chalcogenide							
Component 2: Electrode		Mn: CdSe	CdS-NH_2GO	g-C_3N_4 /CdS	CdS-GOD	CdSe QDs		**Component 4: Method**
	ITO/Bi_2WO_6/CdS	amyloid beta (Aβ)					PEC	
	Au/EDC-NHS		(DENV) E-protein				SPR	
	FTO/Pt/Ag			U-PA			PEC	
	FTO/ZnO(IOs)				AFP		PEC	
	Au/EDC-NHS/*sig DNA*					Streptavidin	EC	
		PBS	Au/IgM	BSA, CS (chitosan)	BSA (blocking agent), CS	BSA (bovine serum albumin)		
	Component 3: Supplement							

Notes: Detectable protein by biosensor is written in the cell that is the confluence place of four table components. Vacant cells are proposed for other probable protein detection.

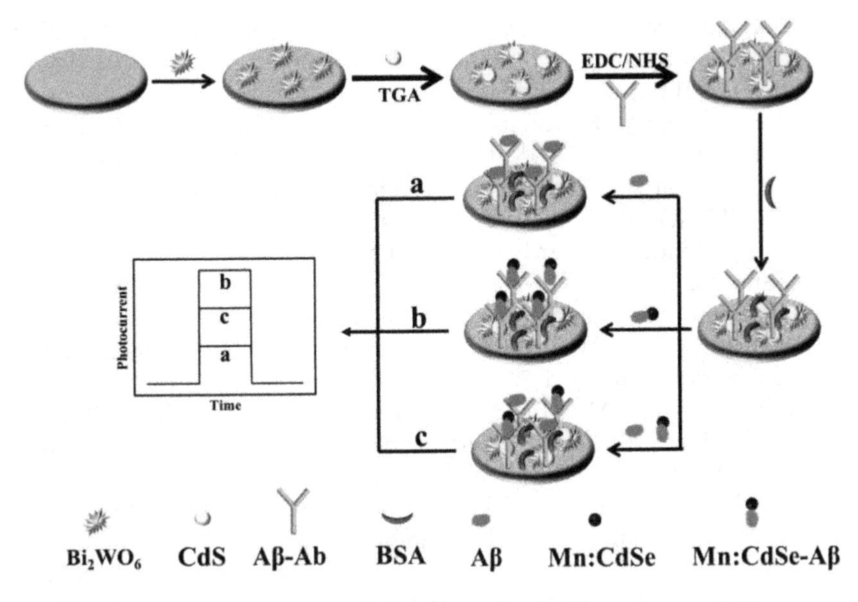

FIGURE 7.2 Fabrication process of photoelectrochemical immunosensor [17].

flower-like of Bi_2WO_6 provided a large specific surface for loading functional CdS. This design leads to excellent photocurrent, enabling the sensor to fabricate. Mn: CdSe was used to label Aβ for acquiring Mn: CdSe-Aβ bioconjugate. By this action, the tracing sensitivity is enhanced via the competitive immunoreaction of Mn: CdSe-Aβ and Aβ with antibody immobilized on the Bi2WO6/CdS electrode. The doped of Mn^{2+} in CdSe nanoparticles could induce energy defects that prevented the combination again of photo-generated charges. Therefore, the PEC response for ultrasensitive detection of Aβ was promoted. The fabricated immunosensor for Aβ showed a linear range of 0.2 pg/mL–50 ng/mL with a detection limit of 0.068 pg/mL, also with good stability, high selectivity, and acceptable reproducibility. The sensitization of Mn: CdSe furnished a novel route for the fabrication of highly sensitive protein biosensors.

Fig. 7.2 represents that the Aβ antibody was gathered on Bi2WO6/CdS and improved ITO electrode to prepare of Aβ (1~42) PEC immunosensor. Also depicts different concentrations of target Aβ and Mn: CdSe-Aβ mixture was incubated to accomplish a competitive immunoreaction.

Dengue virus (DENV) is an infected disease in tropical areas that have encountered human life with danger. Omar and coworkers developed an optical sensor with the goal of the dengue virus (DENV) E-protein detection. The biosensor functioned based on cadmium sulfide quantum dots composited with amine-functionalized graphene oxide (CdS-NH_2GO) thin film. To measure targeted E-proteins, specific monoclonal antibodies (IgM) were covalently appended to CdS-NH2GO via EDC/NHS coupling. The SPR sensor exhibited an excellent detection limit (0.001 nM/1 pM) with a sensitivity of 5.49°/nM for the measurement of DENV E-protein. The binding affinity, as well as the performance of the Au/CdSNH_2GO/ EDC-NHS/IgM film, was successfully obtained at 486.54 nM^{-1} in recognizing DENV E-proteins. The Au/CdS-NH_2GO/EDC-NHS/IgM film shows high potential sensitivity and stronger binding towards DENV E-protein [18].

Wei and colleagues [19] established a biosensor based on a steric hindrance hybridization assay to allow the highly sensitive detection of streptavidin. In the steric hindrance hybridization measurement, three labelings were performed. These 3 labelings include: the signaling strand DNA (sig-DNA) at the 3′ end with CdSe quantum dots (QDs), at the 5′ end with biotin, and capturing strand DNA (the complementary strand of sig-DNA) at the 5′ end with thiol. streptavidin produced the steric hindrance effect, which was bound with the signaling DNA strand. The streptavidin limited the ability of the sig-DNA to hybridize with the cap-DNA, which were bonded on the surface of the Au electrode. Hence, streptavidin concentration was measured indirectly based on the CdSe QDs concentration against the electrode surface. The used method in this measurement was differential pulse anodic stripping voltammetry. In the optimal situation, the streptavidin detection range was 1.96 pg/mL to 1.96 μg/mL by the biosensor and the detection limit was 0.65 pg/mL. Based on the experimental conclusion, the as-prepared biosensor has the ability of rapid and accurate streptavidin detection. The perfect fabrication steps for the electrochemical biosensor are demonstrated in Fig. 7.3.

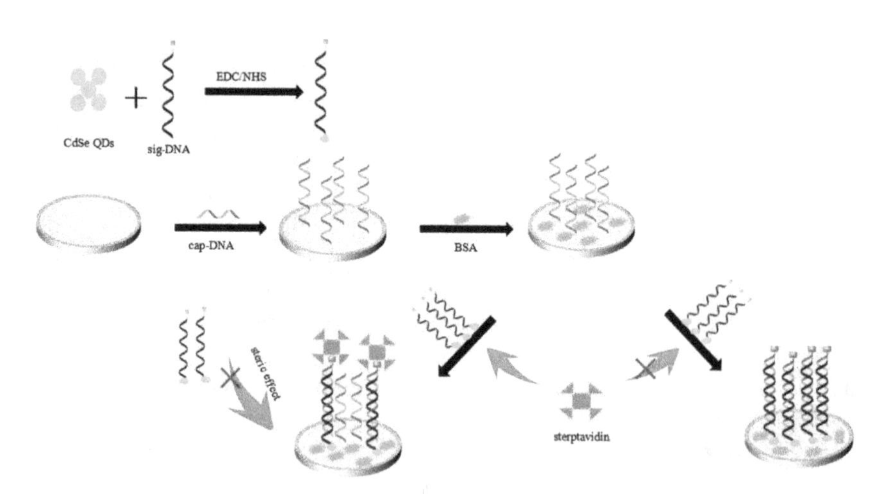

FIGURE 7.3 Fabrication steps for the electrochemical biosensor [19].

Alpha-fetoprotein (AFP) is a major plasma protein that originates from the liver, yolk sac, and fatal human gastrointestinal tract. The probability of symptoms of a malignant tumor is increased when high concentrations of AFP are observed in adult serum. This detection applies as a particular diagnosis of preliminary liver cancer clinical indicators.

The efforts of Xu and teammates [20] lead to biosensor fabrication for AFP detection. In the biosensor, ZnO inverse opal-structure (IOs) based photoelectrochemical (PEC) Electrode and AFP-CdS-GOD (glucose oxidase) composite were used. AFP detection is possible by the competitive immunosensor of AFP and the AFP-CdS-GOD composite with anti-AFP antibodies (Ab) immobilized on FTO (fluorine-doped tin oxide)/ZnO IOs electrode. For hydrogen peroxide(H_2O_2) production, Catalyzation of glucose via GOD is feasible. Electron donor H_2O_2 scavenging photogenerated holes in the valence band of CdS QDs and so reduces from joining electrons to holes of CdS QDs. The effective energy level between the conduction bands of CdS QDs and ZnO broadened the range of light absorption. This broadening allowed electron transition from excited CdS QDs to ZnO upon visible light irradiation, which raised the photocurrent. Based on the results, the immunosensor of AFP bears a large linear detection range of 0.1–500 ng/mL with a recognition limit of 0.01 ng/mL. The stable biosensor is devoid of interference properties and could be an effective nominee for the detection of other proteins.

Urokinase-type plasminogen activator (U-PA), a member of the serine protease family, functions in the human plasminogen activation system. U-PA is critical for tissue degradation and remodeling, cell migration, extravascular fibrinolysis, and cancer invasion and metastasis. Liu and coworkers [21] used g-C_3N_4/CdS nanocomposite for the establishment of a new ultrasensitive U-PA biosensor. The prepared nanocomposite was deposited on the plane of a bare FTO electrode. To produce a U-PA biosensor, the modified electrode was incubated with antibody versus urokinase-type plasminogen activator and the blocking agent BSA. In the presence of the target u-PA antigen, the photocurrent response of the accumulated biosensor electrode decreased significantly. The biosensor exhibited a low detection limit of 33 fg/mL, ranging from 1 μg/mL–0.1 pg/mL.

As demonstrated in Fig. 7.4, bare fluorine-doped tin oxide (FTO) electrode was pre-treated with PDDA, then g-C_3N_4/CdS nanocomposite was deposited on the electrode. A uniform nano-film of the nanocomposite was formed after air-drying. Then the g-C_3N_4/CdS nanocomposite-modified electrode was used as the matrix of the biosensor [21].

Golgi protein-73 (GP73) is a promising marker for monitoring liver tumors. Liu and colleagues [22] developed an improved fluorescence biosensor for GP73, by using an anti-GP73 antibody (GP73 Ab) capped

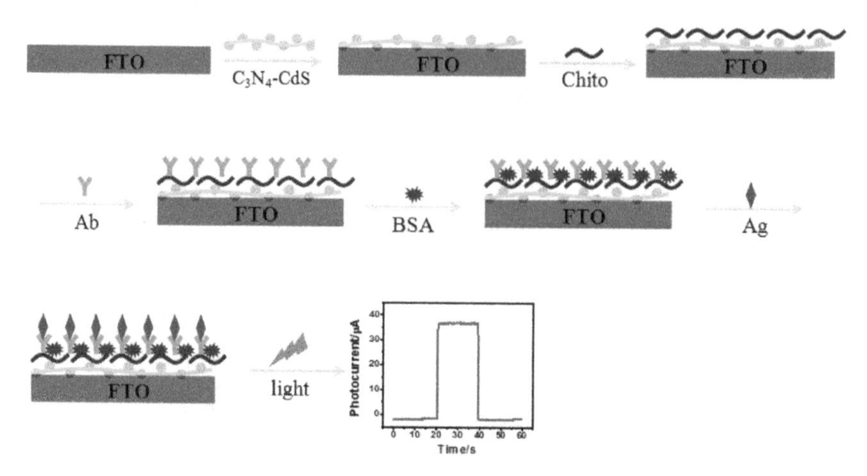

FIGURE 7.4 Schematic diagram of the entire fabrication process for the photoelectrochemical sensor.

quantum dots (QDs) coupled with protein A/G agarose beads. At the first step, carboxylic-functionalized Mn modified CdTe/CdS QDs conjugated with GP73 Ab via a covalent bond. In the second step, protein A/G agarose beads were combined with the QDs-conjugated Ab particularly. After the formation of the QDs-Ab-beads conjugate at this step, the conjugate could capture and separate GP73 from the sample through ordinary centrifugation. In the third step, after GP73 addition, the fluorescence intensity of the above QDs-Ab-beads biosensor could be quenched particularly. Application of the as-prepared QDs-Ab-beads biosensor suggests a simple, rapid, and specific quantitative method for GP73 protein. In optimal conditions, the detection limit of GP73 protein (defined as 3 σ/K) was 10 ng/mL. The fabricated biosensor showed remarkable potential in the clinic test of GP73. Fig. 7.5 exhibits fabrication steps and GP73 detection in summary.

Hu and coworkers [23] succeed in the fabrication of a Cytochrome C (Cyt C) biosensor using graphene oxide sheets/polyaniline/CdSe quantum dots (GO/PANi/CdSe) nanocomposites and electrochemiluminescence (ECL) method. Total process characterization was performed by cyclic voltammogram (CV) and electrochemical impedance spectroscopy (EIS). Experimental parameters such as the ratio of GO/PANi, the $K_2S_4O_8$ concentration, and the pH value of the electrolyte solution were studied to investigate the effect on the ECL intensity. At optimal circumstances, the ECL intensity decreased linearly with the Cyt C concentrations in the range from 5.0×10^{-8} to 1.0×10^{-4} M with a detection limit of 2.0×10^{-8} M. Remarkable specificity, reproducibility, and stability of the fabricated biosensor cause to

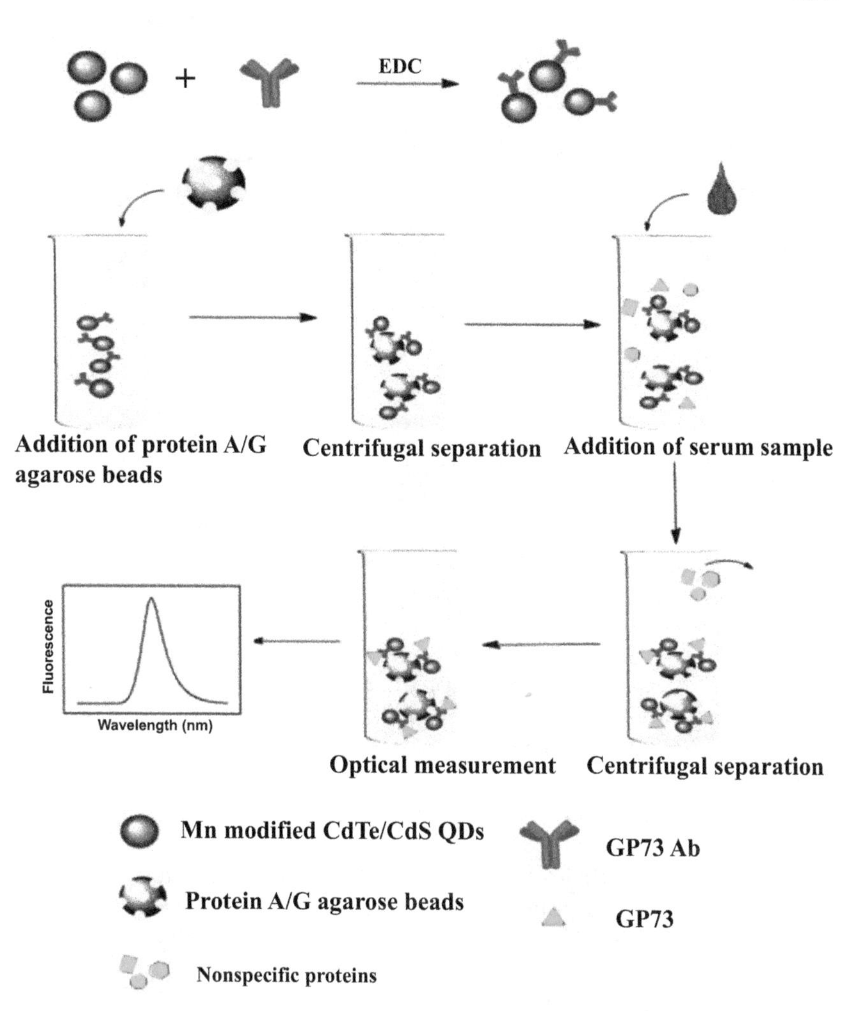

FIGURE 7.5 Schematic of Mn modified CdTe/CdS-GP73 Ab conjugate combined with protein A/G agarose beads to form a nanobioprobe for the detection of GP73 protein [22].

applying in more bioanalytical systems. Fig. 7.6 exhibits the fabrication steps of the ECL biosensor.

- Step 1; Polishing the glassy carbon electrode (GCE, $d = 4$ mm) to a mirror with 1.0, 0.3, and 0.05 μm alumina powder.
- Step 2; Sonication in ethanol and deionized water in turn and drying the electrode at room temperature. Step 3: 10 μL of GO/PANi/CdSe nanocomposites solution was dripped over the surface GCE to construct the GO/PANi/CdSe electrode. Step 4: 3 μL of Nafion solution (0.1 wt.%) was dripped over the area of the GO/ PANi/ CdSe modified electrodes and dried in air.

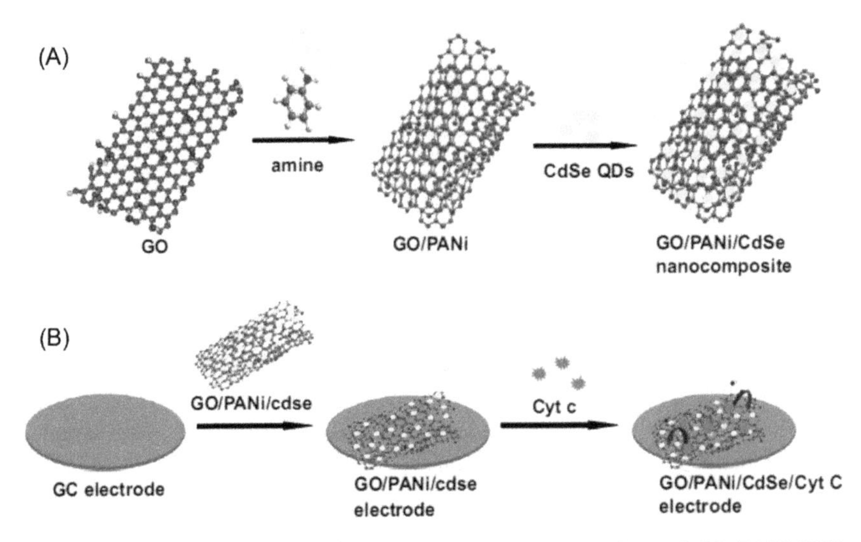

FIGURE 7.6 Schematic representation of preparation procedure of GO/PANi/CdSe nanocomposites (A) and the stepwise biosensor (B).

Hu et al. [24] designed a novel prostate-specific antigen (PSA) biosensor using an "on-off-on" signal switch system combined triple-helix molecular switch with efficient charge separation and transfer between different sensitization units. In this biosensor, the type-II CdTe@CdSe core-shell quantum dots (QDs) labeled on the terminal of the signal transduction probe (STP) moved away from the electrode surface. In this situation, the gold nanoparticles (GNPs) are kept attached close to them. Also, CdS QDs, and ZnO nanotubes, were assembled on Au nanoparticles modified paper fibers. The as-prepared photoelectrochemical (PEC) apt-sensor turned to a "signal-off" state because of the dual inhibition of the disappeared sensitization effect and signal quenching effect of GNPs. Based on the target recognition, the triple-helix structure was perturbed with the formation of the DNA-protein complex and the recovery of the STP hairpin structure, resulting in the second "switch-on" state.

HER2 protein (human epidermal growth factor receptor 2, also called c-erbB2 or Neu) is an important biomarker associated with breast cancer. Xing et al. [25] designed HER2 biosensors by using MnCuInS/ZnS quantum dots (QDs) for their near-infrared (NIR) emission, high quantum efficiency, and low toxicity. MnCuInS/ZnS@BSA nanoparticles as the energy donor for the fluorescence intensity increment formed by encapsulation of MnCuInS/ZnS QDs in BSA. These nanoparticles can easily transfer MnCuInS/ZnS QDs from the organic phase into an aqueous solution. Urchin-like Au nanoparticles (Au NPs) were chosen as the energy acceptor because of their acceptable stability and excellent quenching ability

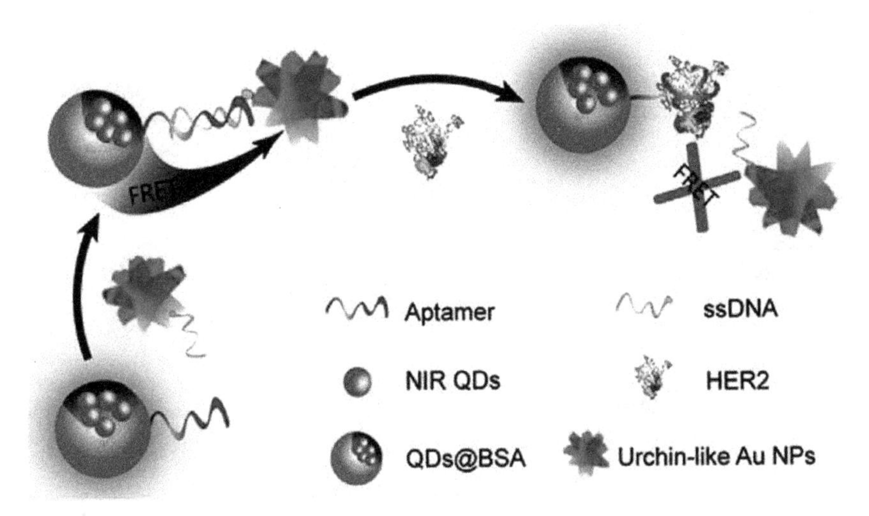

FIGURE 7.7 Schematic exhibition of the enhanced FRET system for HER2 detection based on NIR MnCuInS/ZnS@BSA and urchin-like Au NPs. Note that QDs@BSA is attached to the aptamer, and urchin-like Au NPs are linked to the DNA.

towards NIR fluorescence. The NIR MnCuInS/ZnS@BSA and urchin-like Au NPs were planned as a novel donor-acceptor pair for Fluorescence Resonance Energy Transfer (FRET) measurement. The optical advantages of NIR QDs encapsulated BSA nanoparticles combined with the excellent fluorescence quenching ability of urchin-like Au NPs. The combination led to the as-prepared FRET-based biosensor realizing an enhanced FRET effect for highly sensitive recognition of HER2 in human serum samples. A broad detection range (2e100 ng/mL) and a low detection limit (1 ng/mL) were attained. Therefore, interference reduction of other biomolecules in the NIR region could be achieved by applying the sensing system for other biomarkers (Fig. 7.7).

Amino-terminal pro-brain natriuretic peptides (NT-proBNP) are important biomarkers of Heart failure (HF) because of the long half-life period and high consistency in vitro. Wang et al. designed a "signal-on" electrochemiluminescence (ECL) biosensor for NT-proBNP with "luminophore." Mn-doped ZnAgInS/ZnS nanocrystals deposited in thermally treated UiO-66-NH_2 with mesoporous cages for luminophore preparation. The Stable luminescence increase and the surface defects reduction are achieved by Mn doping and ZnS shell respectively. The luminophore was labeled to diagnose antibody (Ab_2) for ECL immunoassay (Fig. 7.8A). The capture antibody (Ab_1) was immobilized on gold nanoparticles (Au NPs) loaded MoS_2@Cu_2S modified electrode. The detection antibody was labeled with the luminophore and the sandwich NT-proBNP biosensor provided. MoS_2 nanosheet deposition on Cu_2S

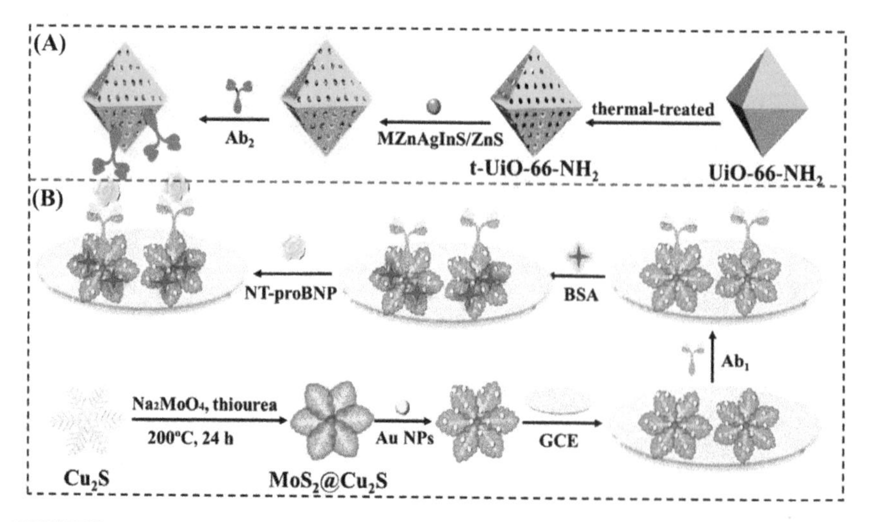

FIGURE 7.8 (A) Preparation of luminophore and Ab_2 bioconjugates; (B) fabrication of ECL immunosensor and its recognition to NT-proBNP.

snowflake was performed to produce the co-reaction accelerator MoS_2@Cu_2S (core-shell) with considerable catalytic active sites (Fig. 7.8B). The snowflake-like core-shell-Au architecture loaded capture antibody (Ab_1) because of its high specific surface area. At optimized conditions, NT-proBNP detection exhibited linear scope of 1 fg/mL–100 ng/mL and a detection limit of 0.41 fg/mL [26].

Huang et al. [27] invented signal-on-off type photoelectrochemical (PEC) biosensors for sensitive detection of carcinoembryonic antigens (CEA).In this biosensor were used tin disulfide (SnS_2)nanosheets loaded on reduced graphene oxide (SnS_2-GR) and gold nanoparticles coated on reduced graphene oxide-functionalized copper sulfide (Au@CuS-GR) for photoactive material and signal amplification respectively. The photocurrents of SnS_2-GR were quenched by Au@CuS-GR (signal amplifier). The amplification is accomplished by the p–n type semiconductor quenching effect and the steric hindrance effect. The catalytic performance of the amplifier as peroxidase mimetics could oxidize 4-chloro-1-naphthol (4-CN) to an insoluble product on the electrode surface. Based on the multiple signal amplification strength of Au@CuS-GR, CEA was detected sensitively with a linear range from 0.1 pg/mL to 10 ng/mL and a limit of recognition down to 59.9 fg/mL (S/N = 3).

Another CEA biosensor was devised by Zhang et al. [28] using MOF-derived porous hollow carbon nanobubbles@ZnCdS multi-shelled dodecahedral cages (C@ZnCdS MSDCs). Remarkable label-free PEC performance of C@ZnCdS MSDCs arising from synergistic effect between the porous shells and the carbon-layer coating. For type-II heterostructures formation,

TiO_2 improved ITO electrodes then C@ZnCdS MSDCs were synthesized and assembled onto the electrode. Afterward, Au NPs were deposited on the ITO/TiO_2/C@ZnCdS MSDCs area. At optimum circumstances, the linear range of the biosensor is broad (0.00005–500 ng/mL), and the detection limit are low (2.28 Fg/mL). This PEC biosensor lightens new capabilities for the application of photoactive material and MOFs-derived hollow carbon materials in sensing.

The β-casomorphin 7 (BCM-7) is a typical biomarker of autism disorder. Roushani and colleagues [29] designed an ultrasensitive electrochemical apt-sensor for BCM-7 measurement. At the first step of the biosensor fabrication, thiourea capped-ZnS QDs (ZnS-QDs) and AuNPs modified a glassy carbon electrode (GCE) to immobilize the amino-aptamer (NH_2-Apt) on the electrode surface. The Au–N bonds were formed among the AuNPs and amino groups of the NH_2-Apt (as a receptor of the BCM-7). These bonds caused attachment of the Apt to the electrode surface. Acceptable performance of the BCM-7 apt-sensor satisfied by wide linear concentration ranges from 1 fM to 0.6 μM with a limit of detection (LOD) down to 350 aM.

Chang et al. [30] devised a fluorescence turn-on biosensor for glycoproteins determination using boronic acid functional polymer capped Mn-doped ZnS quantum dots (QDs@MPS@AAPBA). Glycoprotein addition into the QDs@MPS@AAPBA system caused fluorescence intensity increment. The fluorescent probe was used in a broad pH range of 5.0–9.0. Based on experimental data, the binding constants and recognition limits of the system were $1.53 \times 10^7\ M^{-1}$ and 3.36×10^{-10} M for transferrin (TRF). Therefore, the as-prepared biosensor could be applied for TRF detection in serum without any complex pretreatment and the recovery was in the range of 95.7%–103.0%.

7.3 Glucose

Glucose is a monosaccharide with the chemical formula $C_6H_{12}O_6$ (Fig. 7.9). Glucose is the most prevalent monosaccharide, a kind of

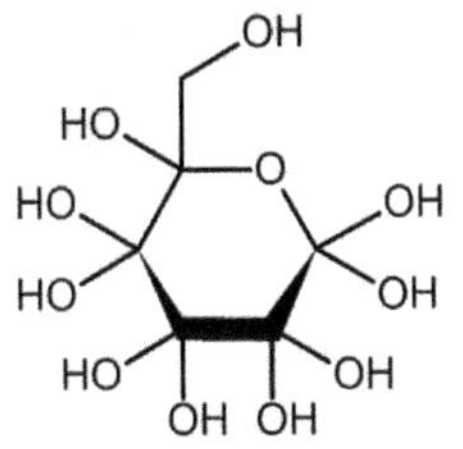

FIGURE 7.9 Glucose molecule.

carbohydrate. Glucose is primarily synthesized by plants and the majority of algae during photosynthesis from water and carbon dioxide with the help of solar energy, where it is utilized to synthesize cellulose in cell walls, the world's most abundant carbohydrate. Glucose is a critical biomolecule found in the body that aids in the physiological generation of the energy molecule adenosine triphosphate (ATP). Typically, glucose is stored in the body by external sources such as nutrition, which varies according to an individual's habits and lifestyle. While glucose is a necessary biomolecule, excessive amounts can result in a variety of metabolic diseases, including diabetes mellitus, as well as a catastrophic failure, including death. Therefore, extreme caution is essential to mitigate the effects of diabetes. As a result, several low-cost approaches for controlling and monitoring glucose levels have been developed, including surface plasmon resonance, optical techniques, colorimetric techniques, and electrochemical techniques [31].

Glucose detection is critical in a variety of industries, including clinical diagnostics, biomedicine, and the food business. Currently, glucose is sensed in circulation by enzymatic glucose sensors that rely on glucose-oxidase to estimate glucose levels via indirect oxidation. Traditionally, the glucose detecting device consisted primarily of a glucose sensor, which was typically electrochemical [32]. In recent years, significant progress has been made toward developing non-enzymatic glucose sensors based on earth-abundant transition metal chalcogenides, transition metal phosphides, transition metal nitrides, transition metal oxides, transition metal hydroxides, transition metal phosphates, and other transition metals [33].

In the light of the glucose sensors, nanostructured materials such as metal nanoparticles, metal oxides, carbon nanomaterials, conducting polymers, and their nanohybrids have had a dramatic influence on electrochemical applications, over the last decade. Transition metallic nanoparticles, such as Au, Pt, Cu, Ni, and Co-based hybrids, are among the nanomaterials that have been employed to increase electrochemical activities [34]. As a result of their outstanding redox behavior, high surface-to-electro-active area ratio, and strong biocompatibility. Recent research has focused on nickel-based chalcogenide for use in various energy applications such as supercapacitors, batteries, hydrogen evolution reactions, and solar cells as well as glucose sensing [31] due to their low cost, layer-dependent physical and chemical properties, as well as their low cost and layer dependent physical and chemical properties.

7.3.1 Chalcogenide-based glucose sensor studies

A high-performance non-enzymatic glucose sensor based on one-step electrodeposited nickel sulfide was investigated by Kannan and Rout

[35]. In the study, A one-step electrodeposition approach was used to generate a nanostructured NiS thin film, and the structural and morphological properties of the films were examined using X-ray diffractometry (XRD), field emission scanning electron microscopy (FESEM), and energy dispersive X-ray analysis (EDAX). The electrocatalytic activity of NiS thin film toward glucose oxidation was explored by manufacturing a non-enzymatic glucose sensor and characterizing its performance using cyclic voltammetry (CV) and amperometry. The constructed sensor demonstrated high sensitivity and a low detection limit of 7.43 $\mu A\ \mu M^{-1}\ cm^{-2}$ and 0.32 μM, respectively, with an 8 s reaction time.

On the other hand, the nanostructured material of $NiSe_2$ nanosheets ($NiSe_2$-NS) has been synthesized by a facile hydrothermal method for electrocatalytic applications [31]. To further develop the electrochemical sensing application of selenide-based metal chalcogenide, the produced $NiSe_2$-NS was employed as a modified electrode material for the electrochemical oxidation of glucose to further develop electrochemical sensing applications. Thus, the electrochemical characteristics of $NiSe_2$-NS with respect to glucose oxidation were determined using cyclic voltammetry and the amperometric i-t method, respectively. Because of the increased electro-active surface area, conductivity, and stability of the newly discovered $NiSe_2$, it exhibits outstanding electrocatalytic activity for the detection of glucose. Comparing the acquired electrochemical parameter values to previously reported modified electrodes, we found that LOD (23 nM), sensitivity (5.6 $\mu A\ \mu M^{-1}\ cm^{-2}$), and linear range (0.099–1252 μM) were all much higher. As a result, the disclosed glucose sensor has a lower detection limit than existing Ni-based modified electrodes, which is advantageous. The fabricated $NiSe_2$-NS/GCE demonstrates exact recovery values, indicating that it has the potential to be used in real-time glucose detection in real-time sample analysis applications.

Copper selenide is a metal chalcogenide that has recently acquired popularity due to its electrocatalytic properties [36]. Copper selenide is composed of CuSe, Cu_2Se, Cu_3Se_2, Cu_7Se_4, Cu_5Se_4, Cu_2Se, and $Cu_{2x}Se$. In a study by Singh et al. [33]. Cu_2Se was identified as a highly efficient bifunctional sensor capable of co-detection of glucose and dopamine with good selectivity after being made through one-pot hydrothermal synthesis. Copper selenide produced in this manner is capable of electrooxidizing glucose and dopamine at a variety of applied potentials. At 0.35 V, glucose oxidized, whereas dopamine oxidized at 0.2 V. This non-enzymatic copper selenide sensor demonstrated great sensitivity for glucose (15.341 $mA\ mM^{-1}\ cm^{-2}$) and dopamine (12.43 $\mu A\ \mu M^{-1}\ cm^{-2}$) with a low limit of detection (0.26 μM and 84 nM, respectively). This sensor's excellent sensitivity and low limit of detection make it an interesting candidate for detecting glucose/dopamine in physiological bodily fluids with low concentrations of these biomolecules. Its extremely low applied potential for detection also makes it excellent for integration into low-power continuous monitoring systems. This sensor demonstrated good repeatability, reusability, and long-term operating stability in the presence of additional interferents, as well as a high degree of selectivity for dopamine and glucose sensing.

7.4 Urea

As discussed in previous chapters, transition metal chalcogenides (TMCs) have been widely investigated for numerous applications ranging from optoelectronics to sensors and solar cells. In some studies, TMCs are used for fabricating wearable sensing devices. Metal chalcogenides can be anchored with active surfaces such as graphene to form metal bounded chalcogenides (MBCs) with improved properties, particularly in the fabrication of sensing devices, where the rate of heterogeneous electron transfer kinetics and the stability of the material are two important factors [37]. Two-dimensional (2D) nanomaterials are renowned electrocatalytic materials, therefore these materials can be used in electrochemical sensors for the detection of various biomolecules such as urea, glucose, etc. TMCs plus carbon materials such as graphene oxide (GO) and carbon nanotubes (CNTs) have shown enhanced kinetic and thermodynamic performances for urea oxidations. MoS2/graphene, MCNS/rGO, and NiCoS4/graphitic nanofibers are some examples in this field. The synergistic effect between GO and TMCs can be observed with higher catalytic properties [38].

Safitri et al. [39] devised a urea biosensor applying a metal chalcogenide (ZnS). In the system, the immobilized urease on the highly luminescent ZnS quantum dots (QDs) functioned as the pH fluorescent label. Amide bond bio-conjugated ZnS QDs to urease and ZnS QDs-urease bioconjugate produced. The immobilized urease caused enzymatic hydrolysis of urea and so pH altered to a more alkaline condition. The alkaline situation deprotonated ZnS QDs, and therefore the fluorescence intensity increased. Linear change of the fluorescence intensity was observed for the urease conjugated ZnS QDs in the urea concentration range of 4×10^{-9}–4×10^{-3} M and pH 6 with a calibration sensitivity of 179.46 intensity/decade. The Ca^{2+}, Mg^{2+}, K^{+}, and Na^{+} ions did not affect the response of the fluorescent pH bio-probe. The urea biosensor was applied for a soil sample and the results showed close matching with the standard DMAB (p-dimethyl amino benzaldehyde) UV-Vis spectrophotometric method. The strategy of ZnS QDs-urease bioconjugation demonstrated broadening the dynamic range and lowering the detection limit for urea in comparison to free enzyme and QDs in solution for urea measurement.

Uric acid detection in human biological samples (like human urine) is possible by non-invasive and simple methods without drawing blood or piercing the skin. Azmi et al. devised a novel fluorescence biosensor based on sol-gel encapsulated CdS quantum dots (QDs)-Uricase/Horseradish Peroxidase (HRP) enzymes in 96-well microplate format. The fluorescence signal of the system resulted from the enzymatic reaction of uricase/HRP in the presence of uric acid. A fluorescence indicator (QDs) revealed the fluorescence signal. When uric acid was added to the hybrid QDs-uricase/HRP, the Allantoin, Carbon dioxide, and Hydrogen peroxide were produced. The QDs fluorescence intensity could be quenched by the created Hydrogen peroxide. Therefore, the intensity is proportional to the uric acid concentration. The main advantage of the as-prepared microplate biosensor is detecting 96 samples per assay within 20 min simultaneously. The linear calibration curve against uric acid was in the concentration range of 60–2000 μM with a recognition limit of 50 μM. [40]. Yang and coworkers [41] designed a new electrode modifier for the simultaneous determination of dopamine (DA), ascorbic acid (AA), and UA. The electrode included hexadecyl trimethyl ammonium bromide (CTAB) functionalized rGO/ZnS (CTAB/rGO/ZnS). The exceptional surface area, good biological compatibility, electricity, stability, high selectivity, and sensitivity of the biosensor owed to combined "reduced graphene oxide" (rGO) and ZnS (as metal chalcogenide) nanoparticles. The linear calibration plots for AA, DA, and UA were acquired over the range of 50.0–1000, 1.0–500, and 1.0–500 μM with detection limits of 30, 0.5, and 0.4 μM, respectively.

7.5 Summary

Metal chalcogenides-based sensors can be used as promising materials in biological detections. Their oxidation activities are greater than other solid-state materials, particularly when combined with materials with a large surface area such as carbon materials. It seems the structural and morphological features of metal chalcogenides govern the superior activity of metal chalcogenides. Transition metal chalcogenides are highly active against biological materials such as glucose, urea, etc. The power density reported for TMCs-based biological electrochemical sensors is one of the highest power densities obtained in direct urea fuel cells (DUFC). As a proof-of-concept, the TMCs are also employed as a new fluorescence sensing platform for the highly selective and sensitive detection of DNA (with a detection limit of 50 pM). It seems that ternary chalcogenide nanosheets may also find use in electrocatalysis, supercapacitors, and electrical devices.

Reference

[1] X. Zhu, H. Zhang, C. Feng, et al., A dual-colorimetric signal strategy for DNA detection based on graphene and DNAzyme, RSC Adv. 4 (2013) 2421–2426. Available from: https://doi.org/10.1039/C3RA44033H.

[2] Y. Yang, C. Li, L. Yin, et al., Enhanced charge transfer by gold nanoparticle at DNA modified electrode and its application to label-free DNA detection, ACS Appl. Mater. Interfaces 6 (2014) 7579–7584. Available from: https://doi.org/10.1021/AM500912M/SUPPL_FILE/AM500912M_SI_001.PDF.

[3] K.J. Huang, Y.J. Liu, J.Z. Zhang, Y.M. Liu, A sequence-specific DNA electrochemical sensor based on acetylene black incorporated two-dimensional CuS nanosheets and gold nanoparticles, Sens. Actuators B Chem. 209 (2015) 570–578. Available from: https://doi.org/10.1016/J.SNB.2014.12.023.

[4] S.N. Sarangi, B.C. Behera, N.K. Sahoo, S.K. Tripathy, Schottky junction devices by using bio-molecule DNA template-based one dimensional CdS-nanostructures, Biosens. Bioelectron. 190 (2021) 113402. Available from: https://doi.org/10.1016/J.BIOS.2021.113402.

[5] G. Zan, Q. Wu, Biomimetic and bioinspired synthesis of nanomaterials/nanostructures, Adv. Mater. 28 (2016) 2099–2147. Available from: https://doi.org/10.1002/ADMA.201503215.

[6] H. Li, X. Wang, J. Xu, et al., One-dimensional CdS nanostructures: a promising candidate for optoelectronics, Adv. Mater. 25 (2013) 3017–3037. Available from: https://doi.org/10.1002/ADMA.201300244.

[7] H. Zhang, Z. Sun, H.M. Xu, et al., Improved quality of life in patients with gastric cancer after esophagogastrostomy reconstruction, World J. Gastroenterol. 15 (2009) 3183–3190. Available from: https://doi.org/10.3748/wjg.15.3183.

[8] N.V. Hullavarad, S.S. Hullavarad, P.C. Karulkar, Cadmium sulphide (CdS) nanotechnology: synthesis and applications, J. Nanosci. Nanotechnol. 8 (2008) 3272–3299. Available from: https://doi.org/10.1166/JNN.2008.145.

[9] T. Zhai, Z. Gu, H. Zhong, et al., Design and fabrication of rocketlike tetrapodal CdS nanorods by seed-epitaxial metal-organic chemical vapor deposition, Cryst. Growth Des. 7 (2007) 488–491. Available from: https://doi.org/10.1021/CG0608514/SUPPL_FILE/CG0608514SI20070115_053304.PDF.

[10] S.K. Kulkarni, A.S. Ethiraj, S. Kharrazi, et al., Synthesis and spectral properties of DNA capped CdS nanoparticles in aqueous and non-aqueous media, Biosens. Bioelectron. 21 (2005) 95–102. Available from: https://doi.org/10.1016/J.BIOS.2004.09.004.

[11] H.M. Jaber Al-Ta'ii, V. Periasamy, Y.M. Amin, Electronic characterization of Au/DNA/ITO metal-semiconductor-metal diode and its application as a radiation sensor, PLoS One 11 (2016) e0145423. Available from: https://doi.org/10.1371/JOURNAL.PONE.0145423.

[12] B.R. Aryal, D.R. Ranasinghe, T.R. Westover, et al., DNA origami mediated electrically connected metal—semiconductor junctions, Nano Res. 135 (13) (2020) 1419–1426. Available from: https://doi.org/10.1007/S12274-020-2672-5. 2020.

[13] D.J. Late, B. Liu, J. Luo, et al., GaS and GaSe ultrathin layer transistors, Adv. Mater. 24 (2012) 3549–3554. Available from: https://doi.org/10.1002/ADMA.201201361.

[14] K. Kumar, M. Sharma, DNA nucleobase interaction driven electronic and optical fingerprints in gallium selenide monolayer for DNA sequencing devices, Appl. Surf. Sci. 571 (2022) 151380. Available from: https://doi.org/10.1016/J.APSUSC.2021.151380.

[15] L. Huang, J.G. Lu, Synthesis, characterizations and applications of cadmium chalcogenide nanowires: a review, J. Mater. Sci. Technol. 31 (2015) 556–572. Available from: https://doi.org/10.1016/J.JMST.2014.12.005.

[16] C. Tan, P. Yu, Y. Hu, et al., High-yield exfoliation of ultrathin two-dimensional ternary chalcogenide nanosheets for highly sensitive and selective fluorescence DNA sensors, J. Am. Chem. Soc. 137 (2015) 10430–10436. Available from: https://doi.org/10.1021/JACS.5B06982/SUPPL_FILE/JA5B06982_SI_001.PDF.
[17] R. Xu, D. Wei, B. Du, et al., A photoelectrochemical sensor for highly sensitive detection of amyloid beta based on sensitization of Mn:CdSe to Bi2WO6/CdS, Biosens. Bioelectron. 122 (2018) 37–42. Available from: https://doi.org/10.1016/j.bios. 2018.09.030.
[18] N.A.S. Omar, Y.W. Fen, J. Abdullah, et al., Sensitive surface plasmon resonance performance of cadmium sulfide quantum dots-amine functionalized graphene oxide based thin film towards dengue virus E-protein, Opt. Laser Technol. 114 (2019) 204–208. Available from: https://doi.org/10.1016/j.optlastec.2019.01.038.
[19] Y.ping Wei, X.P. Liu, Cjie Mao, et al., Highly sensitive electrochemical biosensor for streptavidin detection based on CdSe quantum dots, Biosens. Bioelectron. 103 (2018) 99–103. Available from: https://doi.org/10.1016/j.bios.2017.12.024.
[20] R. Xu, Y. Jiang, L. Xia, et al., A sensitive photoelectrochemical biosensor for AFP detection based on ZnO inverse opal electrodes with signal amplification of CdS-QDs, Biosens. Bioelectron. 74 (2015) 411–417. Available from: https://doi.org/10.1016/j.bios.2015.06.037.
[21] X.P. Liu, J.S. Chen, Mao Cjie, et al., A label-free photoelectrochemical biosensor for urokinase-type plasminogen activator detection based on a g-C3N4/CdS nanocomposite, Anal. Chim. Acta 1025 (2018) 99–107. Available from: https://doi.org/10.1016/j.aca.2018.04.051.
[22] W. Liu, A. Zhang, G. Xu, et al., Manganese modified CdTe/CdS quantum dots as an immunoassay biosensor for the detection of Golgi protein-73, J. Pharm. Biomed. Anal. 117 (2016) 18–25. Available from: https://doi.org/10.1016/j.jpba.2015.08.020.
[23] X.W. Hu, C.J. Mao, J.M. Song, et al., Fabrication of GO/PANi/CdSe nanocomposites for sensitive electrochemiluminescence biosensor, Biosens. Bioelectron. 41 (2013) 372–378. Available from: https://doi.org/10.1016/j.bios.2012.08.054.
[24] M. Hu, H. Yang, Z. Li, et al., Signal-switchable lab-on-paper photoelectrochemical aptasensing system integrated triple-helix molecular switch with charge separation and recombination regime of type-II CdTe@CdSe core-shell quantum dots, Biosens. Bioelectron. (2020) 147. Available from: https://doi.org/10.1016/j.bios.2019.111786.
[25] H. Xing, T. Wei, X. Lin, Z. Dai, Near-infrared MnCuInS/ZnS@BSA and urchin-like Au nanoparticle as a novel donor-acceptor pair for enhanced FRET biosensing, Anal. Chim. Acta 1042 (2018) 71–78. Available from: https://doi.org/10.1016/j.aca.2018.05.048.
[26] C. Wang, L. Liu, X. Liu, et al., Highly-sensitive electrochemiluminescence biosensor for NT-proBNP using MoS2@Cu2S as signal-enhancer and multinary nanocrystals loaded in mesoporous UiO-66-NH2 as novel luminophore, Sens. Actuators B Chem. (2020) 307. Available from: https://doi.org/10.1016/j.snb.2019.127619.
[27] D. Huang, L. Wang, Y. Zhan, et al., Photoelectrochemical biosensor for CEA detection based on SnS2-GR with multiple quenching effects of Au@CuS-GR, Biosens. Bioelectron. (2019) 140. Available from: https://doi.org/10.1016/j.bios.2019.111358.
[28] X. Zhang, J. Peng, Y. Song, et al., Porous hollow carbon nanobubbles@ZnCdS multi-shelled dodecahedral cages with enhanced visible-light harvesting for ultrasensitive photoelectrochemical biosensors, Biosens. Bioelectron. 133 (2019) 125–132. Available from: https://doi.org/10.1016/j.bios.2019.03.028.
[29] M. Roushani, S. Farokhi, F. Shahdost-fard, Determination of BCM-7 based on an ultra-sensitive aptasensor fabricated of gold nanoparticles and ZnS quantum dots, Mater. Today Commun. (2020) 23. Available from: https://doi.org/10.1016/j.mtcomm.2020.101066.

[30] L. Chang, H. Wu, X. He, et al., A highly sensitive fluorescent turn-on biosensor for glycoproteins based on boronic acid functional polymer capped Mn-doped ZnS quantum dots, Anal. Chim. Acta 995 (2017) 91–98. Available from: https://doi.org/10.1016/j.aca.2017.09.037.

[31] S. Mani, S. Ramaraj, S.M. Chen, et al., Two-dimensional metal chalcogenides analogous NiSe2 nanosheets and its efficient electrocatalytic performance towards glucose sensing, J. Colloid Interface Sci. 507 (2017) 378–385. Available from: https://doi.org/10.1016/J.JCIS.2017.08.018.

[32] Y. Zhang, J. Sun, L. Liu, H. Qiao, A review of biosensor technology and algorithms for glucose monitoring, J. Diabetes Complicat. 35 (2021) 107929. Available from: https://doi.org/10.1016/J.JDIACOMP.2021.107929.

[33] H. Singh, J. Bernabe, J. Chern, M. Nath, Copper selenide as multifunctional non-enzymatic glucose and dopamine sensor, J. Mater. Res. Press. (2021) 2–13.

[34] T. Kangkamano, A. Numnuam, W. Limbut, et al., Chitosan cryogel with embedded gold nanoparticles decorated multiwalled carbon nanotubes modified electrode for highly sensitive flow based non-enzymatic glucose sensor, Sens. Actuators B Chem. 246 (2017) 854–863. Available from: https://doi.org/10.1016/J.SNB.2017.02.105.

[35] P.K. Kannan, C.S. Rout, High performance non-enzymatic glucose sensor based on one-step electrodeposited nickel sulfide, Chem. – A Eur. J. 21 (2015) 9355–9359. Available from: https://doi.org/10.1002/CHEM.201500851.

[36] J. Masud, W.P.R. Liyanage, X. Cao, et al., Copper selenides as high-efficiency electrocatalysts for oxygen evolution reaction, ACS Appl. Energy Mater. 1 (2018) 4075–4083. Available from: https://doi.org/10.1021/ACSAEM.8B00746/SUPPL_FILE/AE8B00746_SI_001.PDF.

[37] P. Salarizadeh, M.B. Askari, N. Askari, N. Salarizadeh, Ternary transition metal chalcogenides decorated on rGO as an efficient nanocatalyst towards urea electro-oxidation reaction for biofuel cell application, Mater. Chem. Phys. 239 (2020) 121958. Available from: https://doi.org/10.1016/J.MATCHEMPHYS.2019.121958.

[38] R. Ashwini, V.G. Dileepkumar, K.R. Balaji, et al., Ternary alkali metal chalcogenide engineered reduced graphene oxide (rGO) as a new class of composite (NaFeS2-rGO) and its electrochemical performance, Sens. Int. 2 (2021) 100125. Available from: https://doi.org/10.1016/J.SINTL.2021.100125.

[39] E. Safitri, L.Y. Heng, M. Ahmad, T.L. Ling, Fluorescence bioanalytical method for urea determination based on water soluble ZnS quantum dots, Sens. Actuators B Chem. 240 (2017) 763–769. Available from: https://doi.org/10.1016/J.SNB.2016.08.129.

[40] N.E. Azmi, A.H.A. Rashid, J. Abdullah, et al., Fluorescence biosensor based on encapsulated quantum dots/enzymes/sol-gel for non-invasive detection of uric acid, J. Lumin. 202 (2018) 309–315. Available from: https://doi.org/10.1016/j.jlumin.2018.05.075.

[41] Y.J. Yang, One-pot synthesis of reduced graphene oxide/zinc sulfide nanocomposite at room temperature for simultaneous determination of ascorbic acid, dopamine and uric acid, Sens. Actuators B Chem. 221 (2015) 750–759. Available from: https://doi.org/10.1016/j.snb.2015.06.150.

CHAPTER

8

Fabrication of sensors

8.1 Introduction

In recent years a tremendous amount of work has been carried out on the development of various chemical and biological sensors due to new diseases and problems coming into the picture nowadays. The research and development in this field are emerging as a promising solution to identify diseases so that they can be cured. The early detections of the disease and biomolecules are very important for human health monitoring. More accurately, early detection of the problem is useful to control the situation and prevent it. Different types of biosensors such as electrical, optical, electrochemical, chemical, etc. have been developed for the detection of biomolecules and their concentrations in the human body. A biosensor is a device that measures/detects the target molecule (biological molecule) with the help of a bio-recognition element (e.g., enzyme, antibody, nucleic acid, cells, and tissue) by measuring the change in the physical signal (e.g., optical or electrical) [1–3]. The bio-recognition element and the transducer are the two important parts of the biosensor. The bio-recognition elements are very specific for the target molecules while the transducers convert a biochemical response into a measurable effect in the electrical signal (e.g., a change in current, voltage, or resistance) or in the optical signal (e.g., a change in absorption, emission, or scattering). This change in the electrical or optical signal can be quantified as the change in the amount of the target molecule. Biosensors are attractive for pharmaceutical and biomedical analysis due to their numerous features such as being simple to handle, small, cheap, and able to provide reliable information in real-time which make them comparable to or even better than traditional analytical systems. This chapter focuses on the selection of materials for the fabrication of metal chalcogenide biosensors, methods of

Metal Chalcogenide Biosensors
DOI: https://doi.org/10.1016/B978-0-323-85381-1.00008-8

immobilization/coating of thin films over the substrate, film characterization techniques, and some recently reported metal chalcogenides-based biosensors utilizing various optical, electrical, electrochemical, and colorimetric techniques.

8.2 Selection of material

Biosensor devices/probes are fabricated from a variety of materials for ensuring the necessities of reproducibility, biocompatibility, cost, and suitability for large production. The selection of the material for biosensors is one of the important tasks for its performance. For the fabrication of biosensors, various materials such as organic polymers, sol-gel systems, metals, metal oxides, semiconductors, and various conducting composites are used. The chosen material should be biocompatible, mechanically, and chemically stable and the immobilization of biomolecules on it should be easy. The concentration level of the biofluids for disease detection is generally low, therefore, its detection also depends on the selection of a suitable material that can enhance the detection signal. It is well known that the performance of a biosensor depends on the intrinsic characteristics of the material, e.g., structure, morphology, composition, chemical, and optical properties. Metal chalcogenide nanomaterial possesses a large surface area, good conductivity, and excellent biological compatibility which can be used as a signal amplification element in electrochemical and optical biosensors. In addition, metal chalcogenides exhibit many other excellent properties, such as non-toxicity, quantum, and nanometric scale fabrication, excellent catalytic activity, photo-activity, and unique optical properties [4]. Due to these properties, metal chalcogenides have been found very promising materials for numerous optical, electrical, and electrochemical applications including biomedical applications, imaging, and sensing.

Metal chalcogenides are the combination of an electropositive metal cation and chalcogen (sulfides, selenides, and tellurides) anion [5]. Various chalcogenides, metal chalcogenides, and transition metal chalcogenides such as molybdenum disulfide (MoS_2), tungsten disulfide (WS_2), copper sulfide (CuS), zinc sulfide (ZnS), cadmium sulfide (CdS), tungsten diselenide (WSe_2), antimony telluride (Sb_2Te_3), molybdenum ditelluride ($MoTe_2$), and molybdenum diselenide ($MoSe_2$), are used in biosensor applications [6–8] but the most widely used metal chalcogenide for sensing is MoS_2. It is a layered 2-dimensional (2D) material with a structure similar to graphite. It is composed of three atomic layers: the Mo layer sandwiched between two S layers held together by weak van der Waals interactions. Different types of atoms and molecules can be embedded in this material by intercalation methods and hence it finds

TABLE 8.1 Electronic and optical properties of 2D metal chalcogenides.

Properties	MoS_2	$MoSe_2$	WS_2	WSe_2
Optical bandgap (eV)	2.0	1.7	2.1	1.75
Exciton binding energy (eV)	0.2–0.9	0.5–0.6	0.5–0.7	0.4–0.45
Monolayer thickness (nm)	0.65	0.70	0.80	0.70
Refractive index at 633 nm	5.0805 + i1.1724	4.6226 + i1.0062	4.8937 + i0.3123	4.5501 + i0.4332

many applications. MoS_2 is a semiconductor material and has characteristics similar to graphene. Apart from sensing, it has been used as/in a catalyst, lubricant, lithium batteries, and supercapacitor. In the case of surface plasmon resonance (SPR) based sensors, it enhances sensitivity when coated over the metal film. This is because of effective charge transfer and large electric field enhancement at the metal/MoS_2 interface. It also protects the metal film from oxidation by inhibiting the penetration of oxygen and water molecules. MoS_2 has been used as a single and multilayer film as well as nanosheets for many applications including sensing. The large surface area in sandwich-like structures provides abundant active sites for binding or attachment of analytes which enhance the operating range of the sensor. Table 8.1 summarizes the electronic and optical properties of some of the 2D metal chalcogenide materials [9]. These materials also find applications in SPR-based sensors. All these materials have a large refractive index and the ability to absorb light energy.

A nanocomposite of metal chalcogenide with graphene has been widely used due to its enormous potential for the detection of chemicals and biomolecules. Graphene having a two-dimensional plane structure forms a platform for loading many different particles and provides new pathways for the synthesis of functional nanocomposites with different catalytic, magnetic, and optoelectronic properties. In many nanocomposites, there is a strong synergistic interaction between graphene and its composite components, which can greatly enhance the catalytic activity and stability. The metal chalcogenides used with graphene are sulfides (S), selenides (Se), tellurides (Te), and polonides (Po) but not oxides. These materials are semiconductors in nature with a bandgap typically of 1–3 eV. The nanocomposite of tin sulfide (SnS), one of the binary compounds belonging to the IV–VI group and having a layered structure with graphene has been used for the sensing of Cu^{2+} using the electrochemical method [10]. Graphene-CdS nanocrystals as an immobilization matrix for glucose oxidase have been used for the sensing of glucose [11]. The graphene-CdS nanocomposite film retains the bioactivity of glucose oxidase and increases the stability of the sensor probe for 30 days. Graphene-ZnS nanocomposite along with glucose oxidase has also been used for the fabrication of glucose biosensors. The advantage is that the nanocomposite enhances the electron transfer characteristics of the enzymatic biosensor and retains the biocatalytic activity of the enzyme [12]. An electrochemical sensor based on layered MoS_2-graphene composites has been reported for the detection of acetaminophen [13]. Layered MoS_2–graphene composites have potential applications in third-generation electrochemical biosensors which can facilitate direct electron transfer and enhance the catalytic activity of enzymes. Similar to MoS_2, WS_2 has a layered structure and possesses high catalytic activity. Its composite with graphene has also been used in chemical sensing. Another metal chalcogenide, CuS, has been used in optical materials, solar cell materials, and catalysts. Its application in sensors has also been found as its ultrathin layer offers a large surface-to-volume ratio. It has low conductivity and hence is not suitable for electrochemical sensors. Combining it with graphene improves the charge transfer and its utilization in electrochemical sensors. Apart from graphene, the integration of MoS_2 with conventional materials, such as metal nanoparticles, carbon materials, and conductive polymers has been reported by several groups. Apart from selecting material, for designing biosensors, it is important to study and understand each component that constitutes the probe/electrode as well as all the factors that can affect sensor performance and limitations. After selecting the material the next part is its coating over the substrate. Below various coating techniques have been briefly discussed.

8.3 Coating techniques

For the fabrication of metal chalcogenides-based optical/electrical biosensors, the coating is an important part that can be performed through various techniques such as chemical vapor deposition, physical vapor deposition, sol-gel, dip coating, etc. This section is devoted to the coating process for biosensors applications.

8.3.1 Physical vapor deposition

Physical vapor deposition (PVD) is a thin film coating technique under a vacuum. It is generally used for coatings of pure metals, metallic alloys, and ceramics. As its name suggests, it is the process that involves physically depositing atoms, ions, or molecules of coating material onto a substrate. In this process, first, the material from a solid source is vaporized by a high-temperature vacuum or gaseous plasma and then it is transferred to the substrate surface under vacuum conditions. In the end, the vapor gets condensed onto the substrate and generates a thin film. Nowadays, various PVD systems are available for the deposition of the thin film but thermal evaporation and sputtering are the most commonly used PVD techniques. In the thermal evaporation technique, the coating material is heated to form a vapor that condenses on a substrate to form the coating while the sputtering technique involves the generation of a plasma between the coating material and the substrate. PVD technique is the simplest technique used to prepare amorphous thin films of chalcogenide such as cadmium sulphoselenid (CdSSe) [14], manganese sulfide (MnS) [15], and Ge-Te-Ga [16].

8.3.2 Chemical vapor deposition

It is a method of material synthesis that involves the gaseous reactants in the activated (heat, light, plasma) environment to react with one another and produces a solid film at the surface of the material [17]. In this method, the precursor gases are sent into the reaction chamber at ambient temperatures. As the gases come into contact with a heated substrate, they react or decompose and form a solid phase which is deposited onto the substrate. In the CVD process, the substrate temperature is critical and decides what reactions will take place. This technique is very popular to achieve high-quality materials with wafer-scale or large-scale uniformity, tuneable thickness, and large domain size on varying substrates. The technique is useful for the coating of metal-chalcogenides [18,19]. CdS nanoflakes have been prepared using the CVD method for the sensing of NO_2 [20]. Using this technique, different types of materials with different microstructures can be coated. Some of the disadvantages of this technique are the requirement of high temperature-resistant substrates and ultra-high vacuum.

8.3.3 Dip coating method

Dip coating is a useful method to coat the thin film of polymer or nanocomposite over a substrate. It is a very rich technique for thin film deposition for the cylindrical substrate. In this technique, the substrate is immersed in a vessel that contains the coating material and is removed from it for drying. In the first step, the substrate is immersed in the precursor solution at a constant speed. The substrate is kept in the coating solution for a certain time to get the complete interaction between the substrate and coating material. The time is called the dwell time. After the proper interaction between the substrate and coating solution, the next step is to pull the substrate from the vessel. The pulling speed decides the thickness of the thin film. Apart from pulling speed, the thickness also depends upon the viscosity of the coating solution. By optimizing the pulling speed and the viscosity of the coating material, the desired thickness can be achieved. After pulling the substrate, the excess liquid starts draining from the surface of the substrate. The last step involves the evaporation of the solvent from the substrate and the formation of the thin film. The evaporation process depends on the temperature and the deposition material. The dip coating technique is frequently used for the thin film coating of MoS_2. A uniform large-area three-layer MoS_2 film has been successfully synthesized by dip-coating [21]. In addition, for a chemiresistive-based ammonia sensor, WS_2 nanoflakes have been deposited over alumina substrate using the dip coating method [22].

8.3.4 Spray pyrolysis

Spray pyrolysis is a very simple and cost-effective technique that is used to prepare thin and thick films, ceramic coatings, and powders. The schematic of the spray pyrolysis is shown in Fig. 8.1. The spray pyrolysis equipment consists of an atomizer, precursor solution, substrate heater, and temperature controller. The atomizer in the spray pyrolysis technique is used for the formation of a stream of the liquid which has to be coated over the substrate. The droplet size of the precursor solution depends on the method of atomization. The aerosol spraying produces a larger size droplet while ultrasonic spraying produces smaller size droplets. The substrate is heated over the hot plate for the coating. The deposition temperature and droplet size are the two important parameters to control the surface morphology and thickness

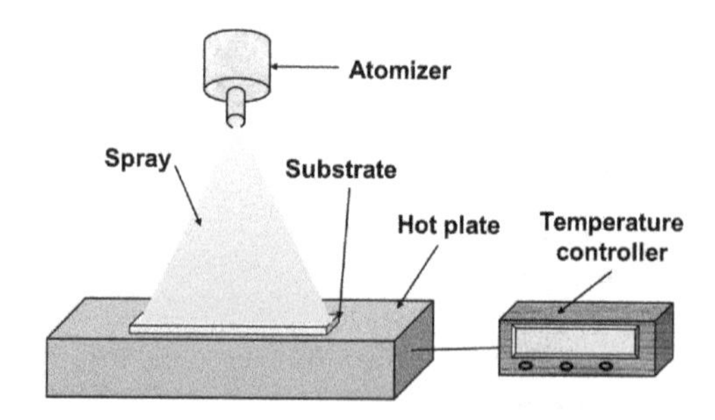

FIGURE 8.1 Schematic illustration of the spray pyrolysis.

of the film. This technique is very frequently used for the thin film deposition of metal oxide, CZTS, and chalcogenides [23]. Semiconducting Cu (In, Ga)(S, Se)$_2$ thin-film is prepared by the spray-pyrolysis method by using the aqueous precursor solutions of copper ($CuCl_2$), indium ($InCl_3$), gallium ($GaCl_3$), and sulfur ($SC(NH_2)_2$) [24]. In addition, the spray pyrolysis method has various advantages such as low-cost setup, simple, rapid film growth rates, the potential for mass production, and reproducibility of the films [25]. Its advantages are low cost, high throughput, and scalable. The disadvantage is the wastage of precursor material during spray coating.

8.3.5 Sol-gel method

Sol-gel is a novel process for the preparation of ceramics and glasses. The film produced by this process is tough, inert, and more resistant than polymer films in aggressive environments. The process involves the transition of a system from a liquid, "sol," into a solid, "gel" phase. Sol is a colloidal suspension of solid particles (1 nm–100 nm in diameter) in a liquid. Colloidal solutions are heterogeneous systems consisting of two phases: a dispersed phase and dispersion media such as water and alcohol. In contrast, the gel is an amorphous porous material containing liquid solvents in the pores. The sol-gel process is often characterized by the nature of the starting liquid used and is used to prepare glasses and ceramics at low temperatures. The process is divided into two steps. The first step is the hydrolysis of a metal alkoxide to produce hydroxide and the second step is the condensation polymerization of a hydroxyl group. Due to this, it is possible to immobilize organic macromolecules in inorganic glass networks.

The starting material (called a precursor) to prepare sol is usually an inorganic metal salt or metal-organic compound such as metal alkoxides having the general formula $M(OR)_x$; M stands for metal with valency x, O is oxygen and R is an alkyl group such as CH_3 (methyl) or C_2H_5 (ethyl) group. An example is tetraethyl orthosilicate (TEOS), and Si $(OC_2H_5)_4$. To perform the hydrolysis reaction of TEOS, it is thoroughly mixed with deionized water and ethyl alcohol. Sometimes, acid catalysts such as hydrochloric acid are also mixed to increase the rate of hydrolysis and condensation. Condensation liberates small molecules, such as water or alcohol. This type of reaction can continue to build up larger and larger molecules through the process of polymerization. A dilute solution containing these polymerized species is coated on the substrate using the dip coating method. For a fixed precursor liquid pH, temperature, pressure, and humidity the film thickness is controlled by the withdrawal speed of the substrate. The process of drying porous material can be divided into several steps. In the first step, the body shrinks by an amount equal to the volume of the liquid that evaporates and the liquid-vapor interface remains at the exterior surface of the body. The second step begins when the body becomes too stiff to shrink and liquid recedes into the interior, leaving air-filled pores near the surface. Eventually, the liquid becomes isolated into pockets and drying can proceed by evaporation of the liquid within the body and diffusion of the vapor to the outside. The conventional method of gel drying is to simply allow the solvent within the gel to evaporate at atmospheric pressure. Gels that have large pores and sturdy pore structures are more easily dried than gels that have fine pores because of capillary forces inversely proportional to the pore size. Drying or evaporation also depends on the liquid solvent used. The advantages of sol-gel coating are high adhesion, ability to coat complex geometries, cost-effectiveness, the possibility of multilayer coating, and no requirement of the conductive substrate. Some of the disadvantages of this technique are the slow rate of coating and film thickness control.

8.4 Synthesis of metal chalcogenide nanosheets

Various methods that have been used for the synthesis of single and multi-layer graphene are also being used for the synthesis of metal chalcogenides. The methods used are both physical and chemical [26,27]. Laser ablation, sputtering, exfoliation by micromechanical cleavage, and ultra-sonication are some of the physical methods. The chemical methods are chemical vapor deposition (CVD), atomic layer deposition (ALD), hydrothermal reactions, and precursor deposition. Intercalation by metals followed by exfoliation in a solvent is used to prepare single and multi-layer chalcogenide films. This has been done for MoS_2 using lithium and water. The requirement is that the material and the solvent should have similar surface energy. The layers prepared are characterized by using many techniques discussed below.

8.5 Film characterization techniques

Various morphological, structural, and elemental characterization tools are employed to gain a better insight into the properties of prepared nanomaterials, and their films and to determine the nanocomposite formation at each synthesis step. This section provides a brief overview of the instruments employed for the characterization of various thin films and nanocomposites along with their fundamental working principle.

8.5.1 Scanning electron microscopy

SEM is used for the morphological analysis of chemically synthesized nanomaterials. In SEM the morphology of a sample is created when a high-energy electron beam (few eV to 50 keV) is impinged on its surface to generate secondary electrons (SE), backscattered electrons (BSE), or characteristic X-rays. The interaction is carried out through a series of electron lenses like in an optical microscope. A pair of condenser lenses collect, collimate and direct the incoming beam towards the objective lens having x, y- scan coils attached to it for providing a faster scan on the sample surface. The reflected beam from the sample is collected by various detectors like an SE detector, a BSE detector, or an energy dispersive X-ray (EDX) detector. The signals are amplified and fed to a cathode ray tube (CRT). The SE detector and BSE detector signals typically provide a black and white morphological image of the sample surface, whereas the EDX detector provides the elemental information in the sample. The beam size and energy determine the resolution of the image and provide a significant depth of field to produce an apparent 3D image of the sample [28–30]. As the working principle is based on a collection of scattered or reflected electrons from the sample, SEM is an advantageous technique in terms of easy sample preparation to obtain the surface morphology with remarkable magnification and clarity. The samples for SEM analysis are prepared by drop casting a uniform dispersion on a glass slide and drying them under appropriate conditions.

8.5.2 Transmission electron microscopy

The more detailed and magnified morphological studies are performed by TEM. TEM provides higher magnification and better resolution as compared to SEM. As the name suggests, in TEM the morphological image is formed by allowing an electron beam to pass through an ultrathin sample and collecting the transmitted beam through different apertures to obtain various kinds of information about the sample. Broadly TEM consists of 4 parts: a high-energy electron source, a condenser lens to focus the electron beam on the sample under investigation, an objective lens that collects the beam transmitted from the sample, and a projector lens that magnifies and projects the sample information on the screen. Certain apertures may be inserted between the objective and projector lens to obtain specific information e.g., selected area electron diffraction (SAED) pattern which provides crystallographic information of the selected area in the sample. Similarly, to obtain the magnified morphology of the sample, two modes of operation may be defined as bright field (BF) and dark field (DF) imaging depending on the direct beam or one of the diffracted beams being allowed to contribute to the image formation at the screen by an aperture at the back focal plane of objective lens [30]. Single and multilayer materials are characterized by TEM.

8.5.3 X-Ray diffraction

The X-ray diffraction method is used to characterize a material to gain insight into its crystallographic phase information. Since the wavelength of X-rays is an order smaller than the visible light, atomic level information may be sought out from a material upon their interaction. Thus, when collimated X-rays are incident on a sample, the periodic array of atoms leads to elastic scattering and the constructive interference of the reflected rays occurs for the angles for which the Bragg's law, $2d\sin\theta = n\lambda$, is satisfied; d is the lattice spacing, λ is the wavelength of X-rays and θ is the angle where constructive interference occurs. For nanomaterials, diffracted X-rays originate from the crystallite planes oriented at a particular angle and hence XRD provides a fingerprint for the identification of its material phase and crystal structure. The detector is coupled with the X-rays source such that it always makes an angle of 2θ. Thus, knowing the values of θ and λ, we may easily obtain the lattice parameter by Bragg's law. Hence, the XRD spectrum provides the peak positions which govern the lattice parameters whereas the peak width gives the information about crystallite size according to Scherrer's equation [31]. A detailed structural study of single-layer MoS_2 carried out by XRD reported Mo atoms coordinated in an octahedral differing from bulk MoS_2 which has trigonal prismatic coordination [32].

8.5.4 Raman spectroscopy

Raman spectroscopy is used to investigate the vibrational modes of a molecule. When laser light strikes a sample it excites electrons of the sample's molecules from one of its vibrational states to virtual excited states to produce in-elastically scattered photons with energy shifted up (anti-stokes lines) or down (stokes lines). This shift in the energy of an incident photon is known as the Raman shift. Thus, a Raman spectrum depicts the scattered intensity of photons as a function of Raman shift, and the peaks manifest characteristics of the molecules. Thus, the Raman spectrum is known as a fingerprint of a molecule [33]. Raman spectroscopy is a versatile tool to study the surface effects of nanomaterials. Samples are simply prepared by drop casting and drying the dispersion of the nanomaterial on a glass slide. Raman spectroscopy can help in the investigation of some of the layer-dependent properties.

8.5.5 FTIR (Fourier transform infrared) spectroscopy

FTIR spectroscopy is employed to study the qualitative bond structure of a molecule for various organic chemicals, polymers, biological samples, etc. Light from an infrared (IR) source is impinged on a sample to investigate the transition between its vibrational levels. A general FTIR spectrometer is typically a Michelson interferometer. Thus, the IR spectrum of a sample is recorded in the form of an interferogram using a Michelson interferometer and then its Fourier transform provides the FTIR spectrum revealing information about the vibrational stretches of a molecule. When IR radiation interacts with the sample, a different amount is absorbed depending on the sample composition. Thus, the peaks or dips in the final spectrograph provide the specific information of bonds in a molecule corresponding to its vibrational motions [34].

8.6 Metal chalcogenides based biosensors utilizing various methods

In recent years, several metal chalcogenides based biosensors have been reported utilizing different methods such as optical, electrical, and electrochemical. In this section recently reported few sensors in each category will be described.

8.6.1 Optical sensors

Optical biosensors reported in the literature utilizing metal chalcogenides are mainly based on SPR and LMR (lossy mode resonance) techniques. These techniques are very powerful and have many advantages. These have been very well described in the literature for the sensing of the number of analytes [35,36]. The main advantage of optical probes over electrical ones is their immunity to electromagnetic interference. To fabricate the sensor either a high refractive index prism or an optical fiber is used as the substrate. Optical fiber has several advantages such as miniaturized probes, low cost, the possibility of online monitoring, and remote sensing. Here, we first describe an SPR-based biosensor that has used a two-dimensional transition metal dichalcogenide and an optical fiber as the substrate for the efficient and rapid detection of bovine serum albumin (BSA) [37]. In this study, the gold-coated unclad core of the optical fiber was coated with MoS_2 nanosheets via the dip coating method. To sense BSA selectively, the MoS_2 surface was bio-functionalized with Anti-BSA antibodies. The importance of the MoS_2 layer was tested by fabricating a probe without the MoS_2 layer and functionalization of the gold surface with Anti-BSA (Ab) antibodies.

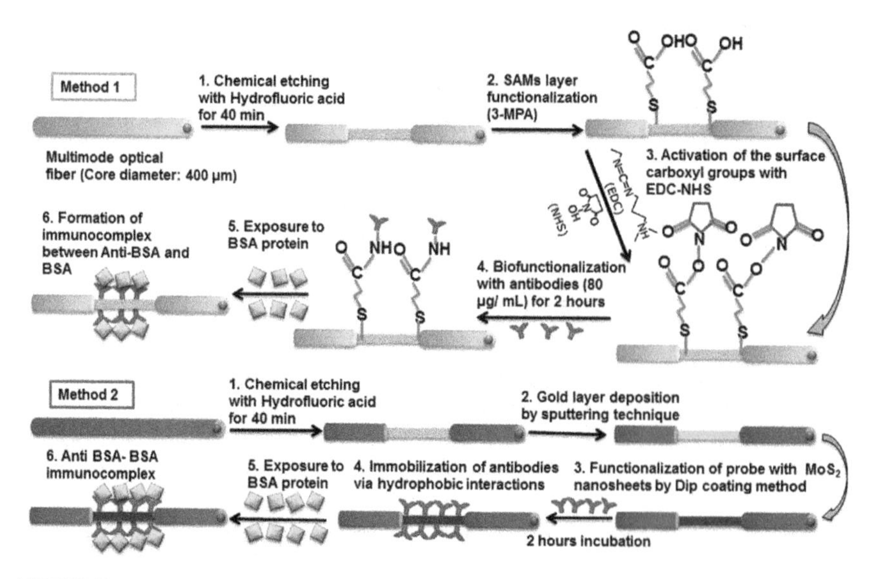

FIGURE 8.2 Schematic of the probe fabrication process of MoS_2-based optical fiber SPR biosensor. *SPR*, surface plasmon resonance. Source: *Figure reprinted from S. Kaushik, U.K. Tiwari, A. Deep, R.K. Sinha, Two-dimensional transition metal dichalcogenides assisted bio-functionalized optical fiber SPR biosensor for efficient and rapid detection of bovine serum albumin, Sci. Rep. 9, 6987 (2019) with permission from Springer Nature.*

In Fig. 8.2, schematics of the BSA probe fabrication method without MoS_2 layer (Method 1) and with MoS_2 layer (Method 2) are shown [37]. Exposure of probe to BSA results in the formation of Anti-BSA and BSA immunocomplex as shown for both the methods in the figure. The formation of complex changes the refractive index of the outer layer called as sensing layer.

The sensor was calibrated using different concentrations of BSA in PBS buffer solution around the probe and recording the SPR spectrum for each BSA sample. Fig. 8.3A and C show the SPR spectra for the probe without (Ab/gold/fiber) and with (Ab/MoS_2/gold/fiber) MoS_2 layer for varying concentrations of BSA sample (pH 7.4), respectively [37]. The results show a shift in the SPR spectrum and hence the resonance wavelength (wavelength of dip) with the increase in BSA concentration around the probes. However, the sensitivity of the probe with a MoS_2 layer is greater than that without a MoS_2 layer as is evident from the slopes of the curves in figures B and D. The slope of the curve (defined as the sensitivity) is more in Fig. 8.3D in comparison to that in Fig. 8.3B [37]. This implies that the addition of the MoS_2 layer plays a very significant role in biomolecule immobilization and sensitivity enhancement of the sensor.

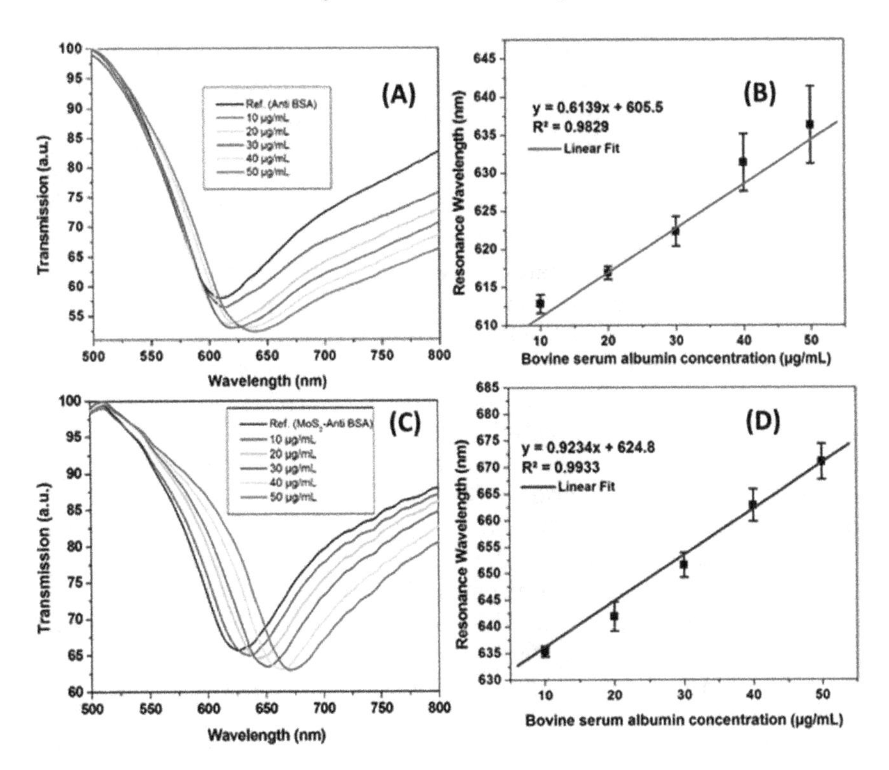

FIGURE 8.3 (A) SPR spectra for different concentrations of BSA sample using probe without MoS_2 layer, (B) variation of resonance wavelength with the concentration of BSA probe for the probe without MoS_2 layer, (C) SPR spectra for different concentrations of BSA sample using the probe with MoS_2 layer, and (D) variation of resonance wavelength with the concentration of BSA using the probe with MoS_2 layer. *BSA*, bovine serum albumin; *SPR*, surface plasmon resonance. Source: *Figure reprinted from S. Kaushik, U.K. Tiwari, A. Deep, R.K. Sinha, Two-dimensional transition metal dichalcogenides assisted biofunctionalized optical fiber SPR biosensor for efficient and rapid detection of bovine serum albumin, Sci. Rep. 9, 6987 (2019) with permission from Springer Nature.*

In the above-described plasmonic sensor, a thin film of gold was coated over the unclad core of the fiber. Now we discuss a plasmonic gas sensor where gold nanohole arrays have been used at the end face of the optical fiber along with layers of a polymer and MoS_2 [38]. The polymer was used because it is a vapor-sensitive material and its electrical and optical properties vary with exposure to vapors at room temperature. MoS_2 was used to facilitate chemical sensing because of its stability, high surface-to-volume ratio, and various attractive sites for the adsorption of gas molecules. The gold nanohole arrays were used for signal-transducing plasmonic nanostructures. The sensor fabricated for methanol gas uses an epoxy (polymer) for its selective detection. It binds methanol molecules via hydrogen atoms and the lone electron pairs on the hydroxyl group. When methanol gas comes in contact with epoxy its molecules diffuse into the epoxy matrix leading to a conformational transformation in its matrix. As a result, its effective refractive index decreases. The decrease in the effective refractive index of the epoxy shifts the plasmonic spectrum and hence the resonance position towards the blue side in the experiments. Apart from this, the local refractive index around the nanoholes increases due to the vapor exposure which shifts the resonance wavelength towards the red side. The measured resonance wavelength shift is, thus, the result of the tradeoff between these two opposite effects. To fabricate the probe gold nanohole arrays were attached at the fiber end face by epoxy adhesive and a MoS_2 layer was coated over the gold nanohole arrays. The sensor structure was abbreviated as Epoxy/Au/MoS_2. The commercial epoxy used serves two purposes: one, as an adhesive to integrate gold nanohole arrays with the optical fiber end face, and two, as the sensitive and selective material. As mentioned, the selectivity of the sensor is based on the decrease in the effective refractive index of the epoxy on interaction with methanol molecules. The high adsorption capacity of MoS_2 concentrates gas molecules locally which enhances the sensor response by a factor of ~2. The sensor is insensitive to relative humidity variation in the range of 11%–92%RH and is highly stable over one year [38].

In addition to experimental studies on plasmonic sensors utilizing metal chalcogenide, some theoretical studies have also been reported. We discuss one such study in which the SPR biosensor structure based on silicon nanosheet and metal chalcogenides has been simulated [39]. The sensor has consisted of a prism, gold thin film, silicon nanosheet, 2D metal chalcogenide film (MoS_2/$MoSe_2$/WS_2/WSe_2), biomolecular analyte layer, and sensing medium. The sensor's performance was evaluated using one of the four chalcogenide materials and the angular interrogation method. Maxwell's equations, Fresnel equations, and transfer matrix method were used to determine the full width at half maximum of SPR curve and sensitivity of the sensor. The simulations were carried out for five different excitation wavelengths. All the models utilizing silicon and metal chalcogenide showed much better performance than the conventional sensing scheme where only Au thin film is used. The highest sensitivity of the sensor was obtained for 600 nm excitation wavelength, 35 nm gold film, and 7 nm silicon nanosheet coated with a monolayer of WS_2 [39].

After discussing SPR-based biosensors utilizing metal chalcogenides we now discuss an LMR-based optical fiber biosensor utilizing the nanocomposite of ZnO/MoS_2 and molecular imprinted polymer for the detection of urinary p-cresol [40]. It is different from SPR in terms of coated layer over the unclad core of the fiber. In the SPR-based sensor unclad core is coated with a noble metal while in the case of LMR the unclad core is coated with a semiconductor metal oxide or a polymer. However, both techniques work on the principle of change in the refractive index of the sensing layer on interaction with analytes. In the SPR-based sensor discussed above, the outer layer was functionalized with Anti-BSA antibodies for the formation of its complex with BSA while in the LMR-based sensor reported in Ref. [40] the bio-recognition layer used is a molecularly imprinted polymer (MIP). MIP provides a highly selective sensing probe and does not require the coating of antibodies or enzymes for selectivity. The molecular imprinting technique involves the preparation of a suitable medium by the polymerization of a monomer in a solvent in the presence of an initiator, cross-linker, and template molecules. The imprints are created with the help of a leaching agent that can break the bond between the functional groups of the polymer and the template molecules. The imprints so created can bind the analyte molecules. For the fabrication of the p-cresol sensing probe, the unclad core of the fiber was coated with a nanocomposite layer of zinc oxide and molybdenum sulfide (ZnO/MoS_2) using the dip coating method followed by the coating of a layer of MIP. ZnO/MoS_2 nanocomposite was prepared by hydrothermal method. The preparation of MIP and the fabrication of the optical fiber probe are shown in Fig. 8.4A and B respectively [40]. Fig. 8.4C shows the experimental setup for the characterization/calibration of the sensor probe. The fabricated probe was fixed in a flow cell used for keeping the p-cresol sample. The LMR spectra were recorded using a spectrometer interfaced with a computer. For the best performance of the sensor following parameters were optimized: p-cresol (template) concentration in the polymer, immobilization period of non-imprinted polymer over ZnO/MoS_2, and the leaching period. The experiments were carried out on p-cresol samples prepared in artificial urine.

Fig. 8.5A shows the LMR spectra obtained for the p-cresol sample concentration ranging from 0 to 1000 μM. From these spectra, the wavelength corresponding to peak absorbance was determined for each concentration. These wavelengths have been plotted in Fig. 8.5B for different concentrations. The peak absorbance wavelength increases with increasing the concentration of the p-cresol due to the interaction of the sample with the imprints. The sensitivity of the sensor determined from the slope of the curve has been plotted in Fig. 8.5C as a function of the concentration of p-cresol. The sensitivity is decreasing due to the decrease in the number of available imprints per p-cresol molecule on the MIP layer as the concentration of p-cresol in the sample is increasing. The performance of the sensor also depends on the pH of the sample and it has been reported to be best for pH 6.0 as evident from Fig. 8.5D.

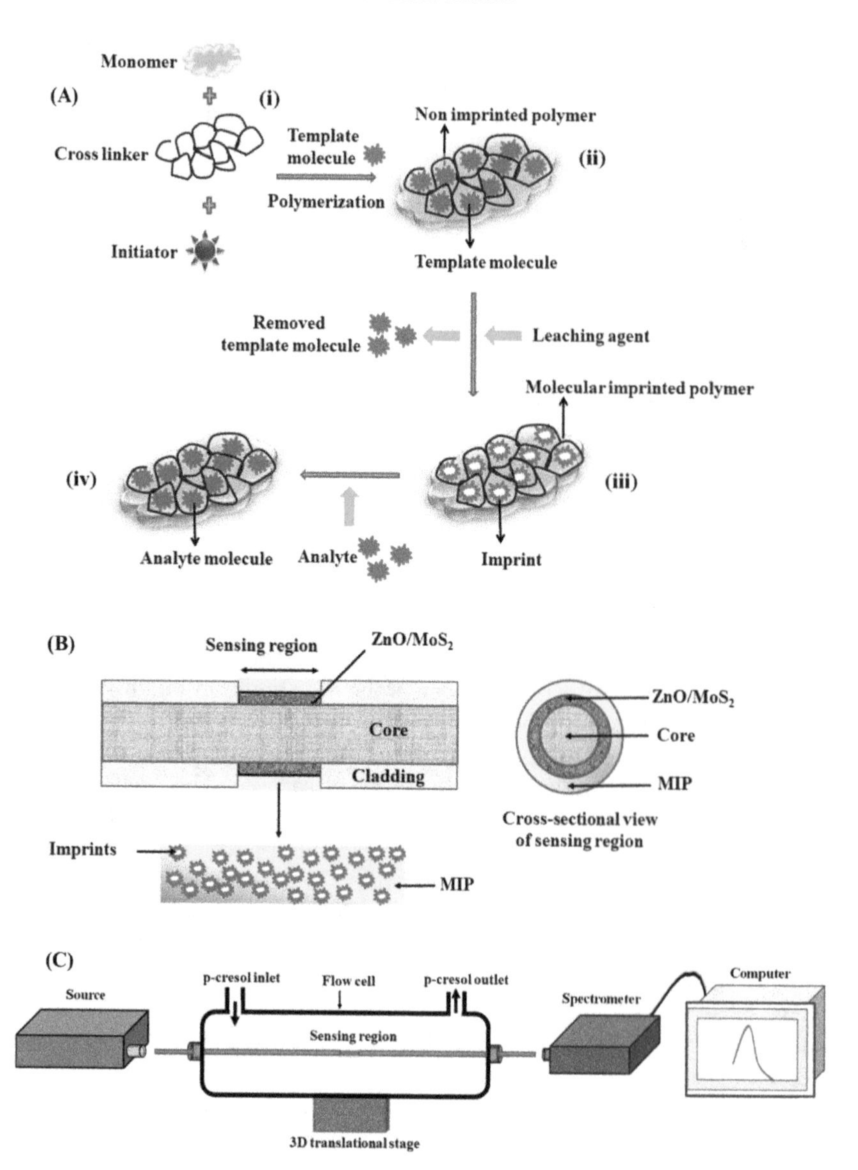

FIGURE 8.4 Schematic of the (A) preparation of molecular imprinted polymer (MIP) and the analyte recognition, (B) fabrication of LMR-based fiber optic probe having layers of nanocomposite of ZnO/MoS_2 and molecular imprinted polymer, and (C) experimental arrangement for the characterization of the probe. Source: *Figure reprinted from S.P. Usha, B.D. Gupta, Urinary p-cresol diagnosis using nanocomposite of ZnO/MoS2 and molecular imprinted polymer on optical fiber based lossy mode resonance sensor, Biosens. Bioelectron. 101, 135–145 (2018) with permission from Elsevier.*

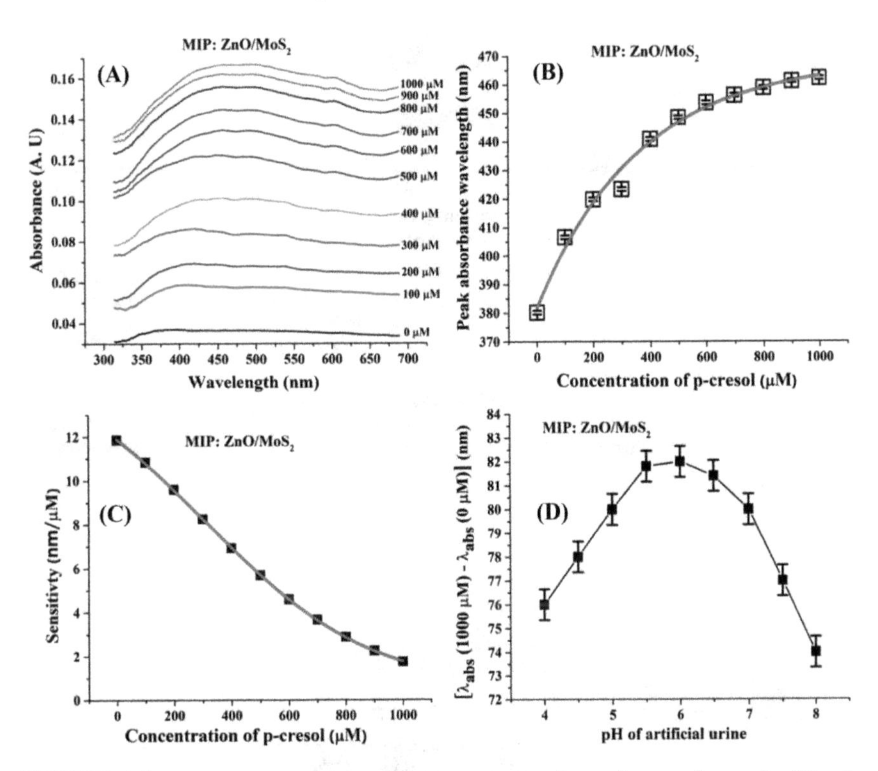

FIGURE 8.5 (A) LMR spectra for different concentrations of p-cresol sample, (B) calibration curve of the sensor, (C) variation of the sensitivity of the sensor with p-cresol concentration, and (D) effect of pH of the sample on the shift in peak absorbance wavelength for the change in concentration from 0 to 1000 μM. Source: *Figure reprinted from S.P. Usha, B.D. Gupta, Urinary p-cresol diagnosis using nanocomposite of ZnO/MoS_2 and molecular imprinted polymer on optical fiber based lossy mode resonance sensor, Biosens. Bioelectron. 101, 135–145 (2018) with permission from Elsevier.*

In addition, the sensor is highly selective, stable, and holds a short response time. In this sensor, MoS_2 in the transducer layer increases the absorption of light in the medium and enhances the LMR properties of zinc oxide due to which the sensitivity of the sensor is enhanced.

An LMR study similar to this has been reported for the detection of creatinine using molecular imprinting over a nanocomposite of MoS_2/SnO_2 [41]. Creatinine is a biomarker to monitor the functioning of kidneys in the human body. The probe was fabricated by depositing a layer of nanocomposite of MoS_2/SnO_2 over the unclad core of the optical fiber as LMR supporting material. It was followed by another layer of MoS_2/SnO_2 nanocomposite along with creatinine imprinted polymer. Thus, in this study, MoS_2/SnO_2 plays a dual role. One, it supports LMR, and two, it gives a platform for MIP to create imprints for the

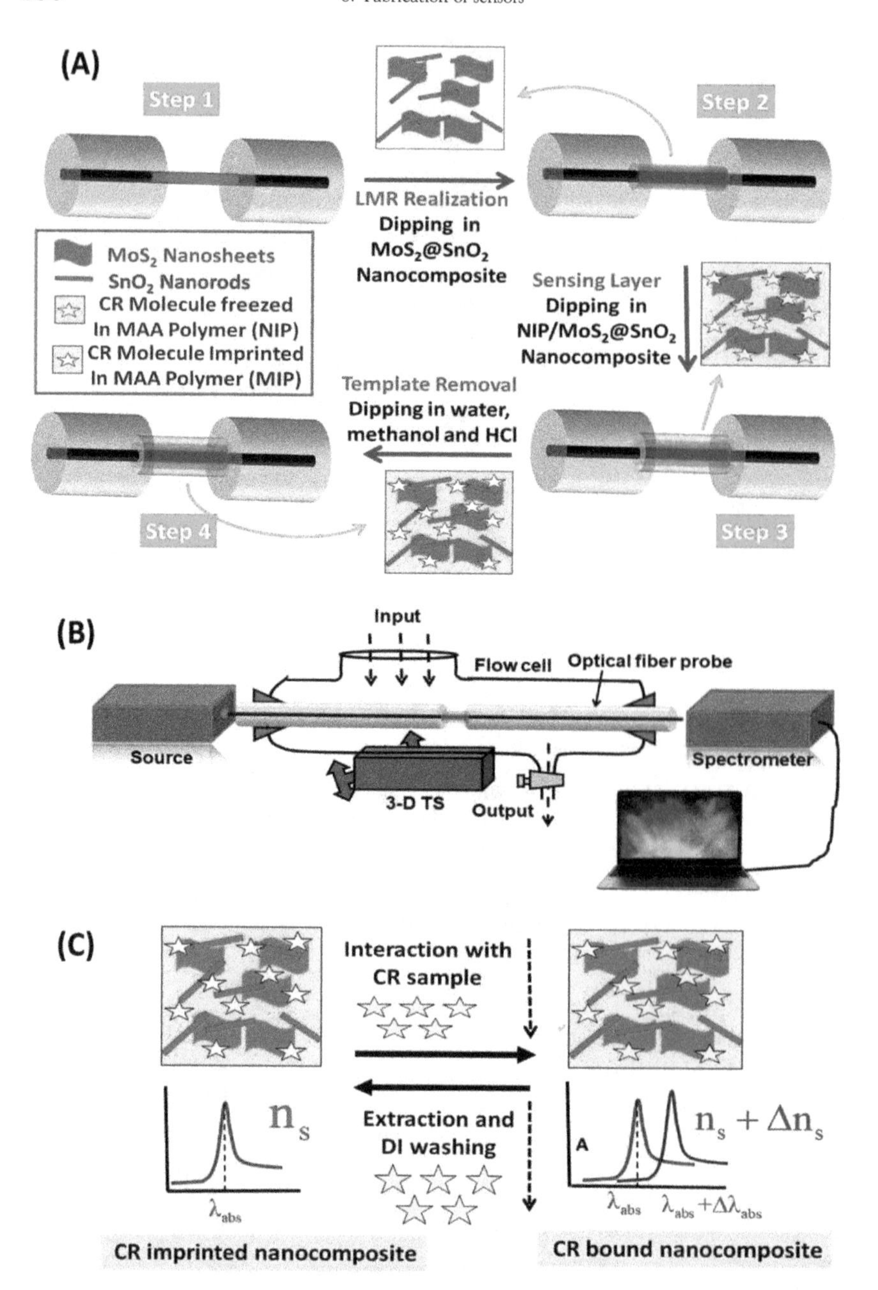

FIGURE 8.6 Schematic of (A) stepwise probe fabrication, (B) experimental setup, and (C) sensing mechanism of creatinine. Source: *Figure reprinted from S. Sharma, A.M. Shrivastav, B.D. Gupta, Lossy mode resonance based fiber optic creatinine sensor fabricated using molecular imprinting over nanocomposite of MoS_2/SnO_2, IEEE Sens. J. 20, 4251–4259 (2020) with permission from IEEE.*

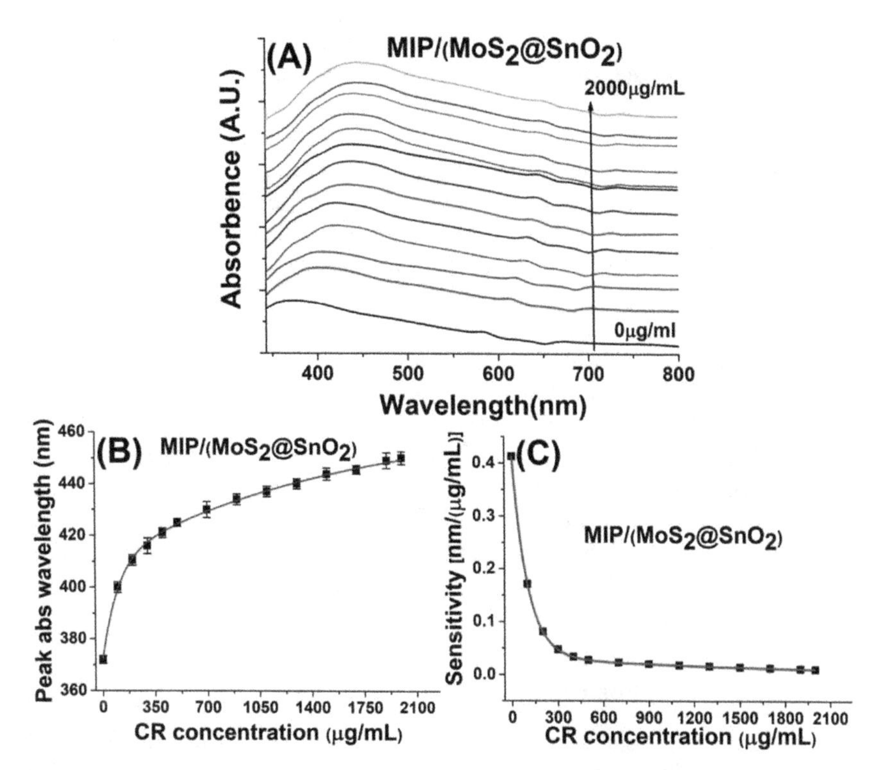

FIGURE 8.7 (A) LMR spectra, (B) calibration curve, and (C) sensitivity for different concentrations of creatinine. Source: *Figure reprinted from S. Sharma, A.M Shrivastav, B.D. Gupta, Lossy mode resonance based fiber optic creatinine sensor fabricated using molecular imprinting over nanocomposite of MoS_2/SnO_2, IEEE Sens. J. 20, 4251–4259 (2020) with permission from IEEE.*

binding of creatinine molecules. The schematic of the probe fabrication is shown in Fig. 8.6A. The advantages of MoS_2 are its great catalytic property to enhance the reaction kinetics, nontoxicity, low cost, environment friendly, and earth-abundance. The molecular imprinting nanocomposite MoS_2/SnO_2 provides specificity to the sensor for creatinine. The sensor's performance has been tested for the creatinine concentration in a physiological range from 0 to 2000 μg/mL using the experimental setup shown in Fig. 8.6B. Fig. 8.6C presents the sensing mechanism of the sensor. LMR spectra, peak absorbance wavelength, and sensitivity for different concentrations of creatinine are depicted in Fig. 8.7A–C respectively. The sensor possesses a maximum sensitivity and detection limit of 0.41 nm/(μg/mL) and 1.86 μg/mL, respectively.

Apart from SPR and LMR, the fluorescence phenomenon has also been used for the fabrication of biosensors utilizing transition metal chalcogenides. The photoluminescence (PL) characteristics of a p-type

WSe_2 monolayer have been studied for optical biosensing applications [42]. The PL characteristics of organic fluorescent dye Cy3 attached p-DNA (probe DNA) coated on the pristine WSe2 (WSe_2/Cy3/p-DNA) were investigated as a function of Cy3/p-DNA concentration. It was observed that an increase in the concentration decreases the PL intensity and shifts its peak to a higher wavelength side suggesting that the WSe_2 layer is p-doped with Cy3/p-DNA. After this, the PL characteristics of WSe_2/Cy3/p-DNA hybridized with t-DNA (target DNA) were investigated. In this case, the PL intensity of the WSe_2/Cy3/p-DNA hybrid system increases, and the peak shifts to the blue side with relatively small amounts of t-DNA (50–100 nM). This occurs due to charge and energy transfer from Cy3/DNA to WSe_2. For t-DNA detection, a system using p-type WSe_2 has merit in terms of an increase in PL intensity. In another study, highly sensitive and selective detection of microRNA has been investigated by combining WS_2 nanosheet-based fluorescence quenching with duplex-specific nuclease signal amplification [43]. The study suggested great potential for this method as a high-performance sensing platform for biomedical and clinical applications.

8.6.2 Electrical sensors

In this section, we shall discuss electrical biosensors that utilize metal chalcogenides. These sensors are based on a field-effect transistor (FET) and MoS_2. One such sensor was reported for label-free detection of pH and biomolecules [44]. In a conventional FET, two electrodes (source and drain) are used to connect a semiconductor material called a channel. A current flowing through the channel between the source and drain is electrostatically modulated by the change in the channel material. For sensing, the surface of the dielectric layer between the source and the drain is functionalized with specific receptors for selectively capturing the desired target biomolecules present around it as shown schematically in Fig. 8.8 [44]. These biomolecules produce an electrostatic effect resulting in the change in electrical characteristics of FET such as drain-to-source current or channel conductance. Apart from the dielectric layer, polymers and lipids have also been used to cover the channel. The only requirement for sensing is the achievement of an electrostatic effect which can take place through dielectric, lipid, polymer, or linker/receptor layers and hence the identification of appropriate channel material. In the first FET biosensor discussed here [44], MoS_2 capable of ultrasensitive and specific detection of biomolecules has been used for the detection of pH and protein. The pH sensing principle is based on the protonation/deprotonation of the OH groups on the dielectric by the pH of the electrolyte which changes the dielectric

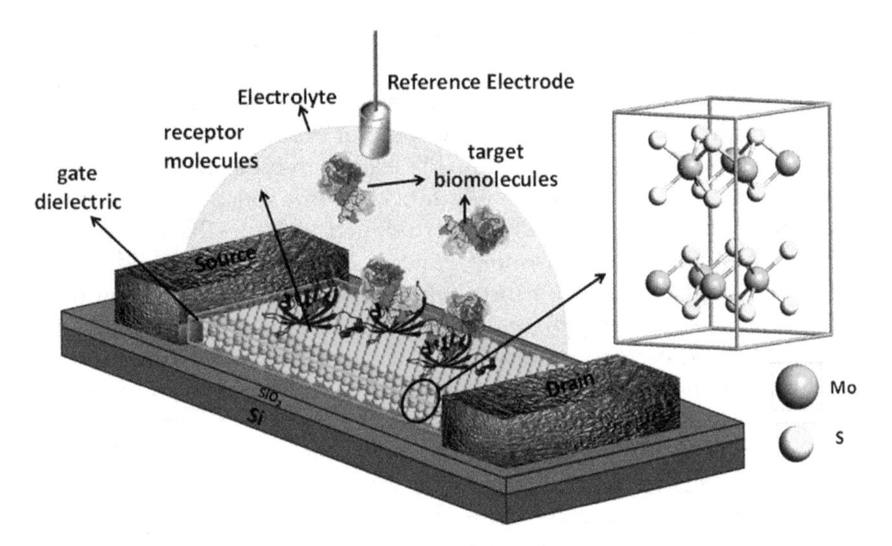

FIGURE 8.8 Schematic diagram of MoS_2-based FET biosensor. *FET*, Field-effect transistor. Source: *Figure reprinted with permission from D. Sarkar, W. Liu, X. Xie, A.C. Anselmo, S. Mitragotri and K. Banerjee, MoS_2 Field-effect transistor for next-generation label-free biosensors, ACS Nano 8, 3992–4003 (2014). Copyright (2014) American Chemical Society.*

surface charge. The low pH of the electrolyte protonates the surface OH groups and generates positive surface charges on the dielectric while the high pH of the electrolyte deprotonates the surface OH groups and generates negative charges. This pH-dependent surface charge determines the effective surface potential and has been used for pH sensing. The MoS_2-based pH biosensor possesses high sensitivity and operates for a pH range of 3 to 9. The sensor has also been shown to detect a specific protein.

Molybdenum disulfide film has also been used to sense ammonia (NH_3) gas [45]. For this study, MoS_2 films were first synthesized by sulfurization of sputtered Mo thin films of 10 nm thickness. The sulfurization produced MoS_2 films of twice the thickness of Mo films. The MoS_2 film and gold electrodes were interdigitated. The sensor performance was evaluated in terms of sensor response defined as relative resistance change. For the measurements, the films were first exposed to a vacuum after loading into the gas sensing chamber. The sensor response was studied at various concentrations of NH_3 from 2 to 30 ppm with a fixed bias voltage. The sensing is based on the following mechanism. When MoS_2 film is exposed to NH_3, the adsorbed gaseous molecules on the surface of MoS_2 shift its Fermi level to a conduction band that decreases the resistance behaving it as an n-type material. The response of the MoS_2 film was very fast and took less than 15 s after the introduction of the gas. The sensor responded linearly to low concentrations of NH_3 [45].

The doping of MoS_2 by metal nanoparticles has also been explored for gas sensing using FET [46]. The doping of noble metals such as Au, Ag, palladium (Pd), and platinum (Pt) lead to p-type material. Out of these noble metals, Pt possesses the highest work function (WF) which causes its highest doping. If Pt NPs are doped in MoS_2 then it can lead to the highest shift in its threshold voltage. In the same way, Pd NPs if doped in MoS_2 can shift their threshold voltage. This fact has been used for the sensing of hydrogen gas using MoS_2 FET with incorporated Pd NPs. If Pd is exposed to hydrogen gas it changes the work function of Pd. Hence, change in the WF of Pd NPs due to the adsorption of hydrogen can be detected in real-time through the change in current in FET [46]. In the MoS_2 FET sensor without the incorporation of Pd NPs, a negligible change in the device current was observed upon exposure to hydrogen (3 ppm). When the same device was used with incorporated Pd NPs, the current level was seen to increase substantially upon exposure to the same hydrogen level as before (without Pd NPs). As mentioned, this is due to the adsorption of hydrogen which led to a decrease in the Pd WF. In both the cases, an appropriate gate voltage was chosen to obtain maximum sensitivity [46].

A label-free biosensor for the detection of cancer marker protein PSA (prostate-specific antigens) has been demonstrated using multi-layer MoS_2 FET with HfO_2—silane-based bio-functionalization of MoS_2 nanosheet for the application of the sensor in liquid-phase [47]. The change in the MoS_2 transistor drain current was found due to the specific binding of PSA to the antibodies that were immobilized over the surface of the MoS_2 film. The sensor has been reported to be highly sensitive and selective not showing a significant response to non-target serum protein [47]. The sensor may find application in the detection of cancer markers at the early stage.

8.6.3 Electrochemical sensors

In this section, we shall describe electrochemical biosensors for the detection of DNA, glucose, and hydrogen peroxide (H_2O_2) utilizing metal chalcogenides. First, we shall discuss the detection of DNA which plays an important role in medical diagnostics, environmental monitoring, drug discovery, and food safety. An electrochemical biosensor for the detection of hepatitis B virus genomic DNA coupled with hybridization chain reaction (HCR) for signal amplification has been reported using tungsten disulfide/multi-walled carbon nanotube composite (WS_2/MWCNTs) [48]. The composite was prepared using a

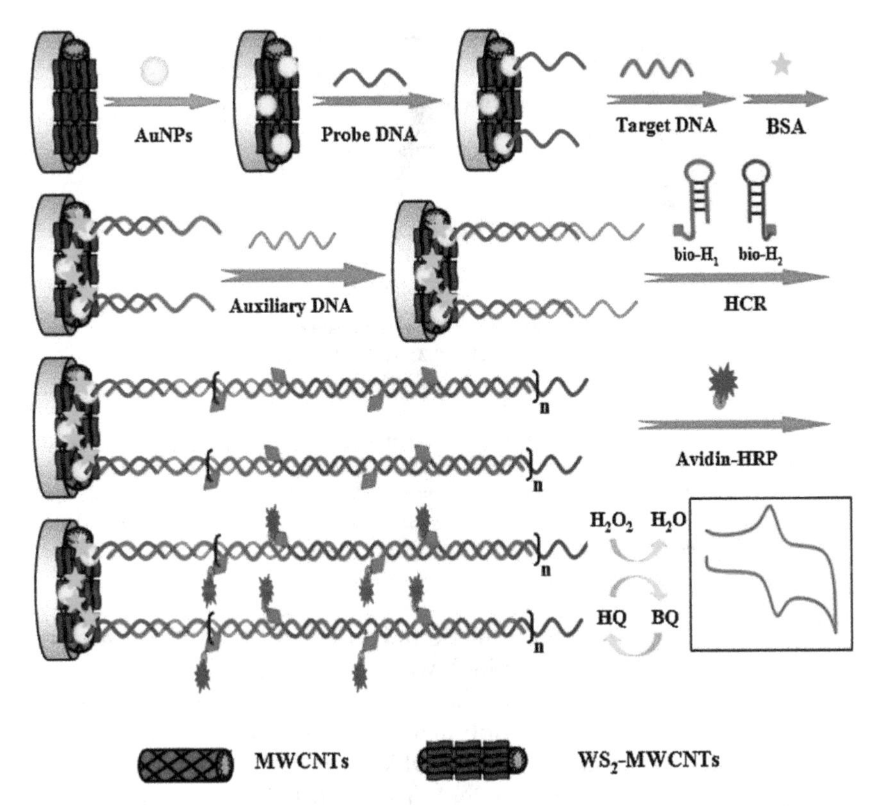

FIGURE 8.9 Schematic of the sensor fabrication with HCR amplification for DNA detection. Source: *Figure reprinted from X. Liu, H.L. Shuai, Y.J. Liu and K.J. Huang, An electrochemical biosensor for DNA detection based on tungsten disulfide/multi-walled carbon nanotube composites and hybridization chain reaction amplification, Sens. Actuators B 235, 603–613 (2016) with permission from Elsevier.*

hydrothermal process. The schematic design of sensor fabrication is shown in Fig. 8.9. For the fabrication of the sensor, first WS_2-MWCNTs were coated over the glassy carbon electrode (GCE), and then electrodeposition of gold nanoparticles (AuNPs) was performed using chronoamperometry. After washing with PBS and drying in the air the electrode was named as AuNPs/WS_2-MWCNTs/GCE. It was then dipped in captured DNA (cDNA) and coated with BSA to name it BSA/cDNA/AuNPs/WS_2-MWCNTs/GCE. Here, BSA blocks the non-specific binding sites. After this, the electrode was first dipped in Target DNA (tDNA) solution and then incubated in auxiliary DNA (aDNA). The obtained aDNA/tDNA/cDNA/BSA/AuNPs/WS_2-MWCNTs/GCE electrode was then dipped in PBS containing bio-H_1 and bio-H_2. After rinsing, the HCR/aDNA/tDNA/cDNA/BSA/AuNPs/WS_2-MWCNTs/GCE electrode was dipped in avidin-HRP.

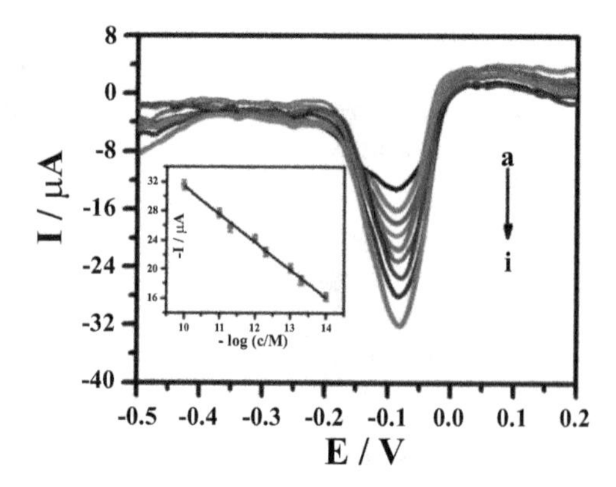

FIGURE 8.10 DPVs of increasing tDNA concentration from a to i. *Inset*: the linear correlation curve. Source: *Figure reprinted from X. Liu, H.L. Shuai, Y.J. Liu and K.J. Huang, An electrochemical biosensor for DNA detection based on tungsten disulfide/multi-walled carbon nanotube composites and hybridization chain reaction amplification, Sens. Actuators B 235, 603–613 (2016) with permission from Elsevier.*

The electrochemical signal of horse radish peroxidase (HRP) catalytic reaction of H_2O_2 and hydroquinol was measured by the differential pulse voltammetry (DPV). For the best performance of the sensor, experimental conditions were optimized. Fig. 8.10 shows the DPVs of different tDNA concentrations. The DPV response increases with the increase in the concentration of tDNA from 10 fM to 0.1 nM and a linear relationship between DPV response and logarithm of the tDNA concentration was observed. The sensor has been reported to have good selectivity, reproducibility, stability, and a remarkably low detection limit (2.5 fM). The synergistic effect of WS_2-MWCNTs composite and HCR greatly improves the sensitivity for DNA detection [48].

In a similar electrochemical sensor reported before this for DNA sensing, MoS_2/MWCNT composite, gold nanoparticle (AuNP), and glucose oxidase (GOD) were used on the electrode [49]. First, the AuNPs were immobilized on MoS_2/MWCNT-chitosan composites modified GCE surface and then GOD was immobilized. The hetero-nanostructure on the electrode accelerates the electron transfer while GOD and AuNPs provide multiple signal amplification which produces ultrasensitive electrochemical detection of DNA. Further, the MoS_2/MWCNT composite film possesses a large specific surface area which increases the immobilization of GOD, but also retains the active immobilized biomolecules and enhances the DNA probe stability. MoS_2 nanoflakes have also been used as electroactive labels for the voltammetric detection of DNA hybridization [50]. This finds application in the diagnosis of Alzheimer's disease. The inherent electroactivity of MoS_2 nanoflakes

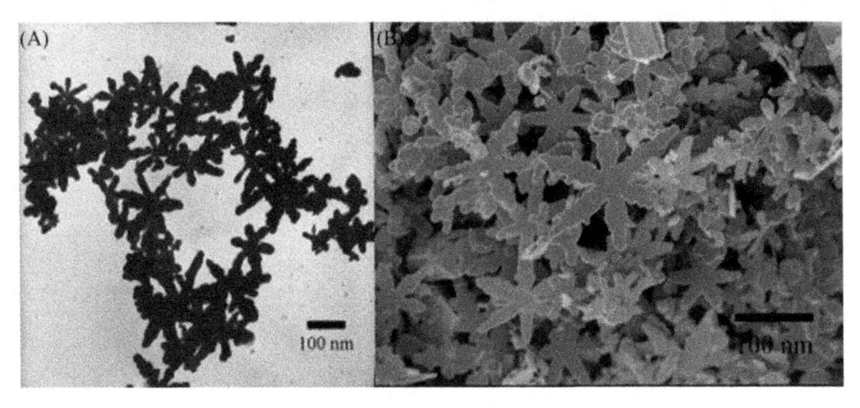

FIGURE 8.11 (A) TEM image of the CuS NPs, and (B) SEM of the CuS/CS/GCE. *TEM*, transmission electron microscopy. Source: *Figure reprinted from Y.J. Yang, J. Zi and W. Li, Enzyme-free sensing of hydrogen peroxide and glucose at a CuS nanoflowers modified glassy carbon electrode, Electrochim. Acta 115, 126–130 (2014) with permission from Elsevier.*

originates from the oxidation of the nanoflakes. In addition, the differential affinity of the nanoflakes towards single and double-stranded DNA is also responsible for this behavior. The range of DNA detection obtained was from 0.03 to 300 nM.

Electrochemical methods have also been used for the enzyme-free detection of hydrogen peroxide and glucose using metal chalcogenides [51]. The sensor was fabricated by modifying GCE with chitosan and CuS nanoflowers synthesized using the hydrothermal method (CuS/CS/GCE). The flower-shaped CuS NPs synthesized can be seen in the TEM micrograph shown in Fig. 8.11A. The SEM image of CuS/CS/GCE has been shown in Fig. 8.11B. The electrocatalytic performance of the bare GCE, CS/GCE, and CuS/CS/GCE toward the oxidation of glucose was investigated by cyclic voltammetry (CV). No current response was observed in the case of bare GCE and CS/GCE. The CV of CuS/CS/GCE has exhibited excellent catalytic capability toward the electro-oxidation of glucose and reduction of hydrogen peroxide in pH 7.2 phosphate buffer. Based on these results, a copper sulfide-based glucose and hydrogen peroxide sensor was developed [51].

The CVs of the CuS/CS/GCE measured for different concentrations of glucose are shown in Fig. 8.12A. As the glucose concentration increases, the peak current of the anodic peak at 0.46 V increases. The variation is linear for the glucose concentration range of 10 μM to 10 mM with a sensitivity of 5.86 μA/mM as shown in Fig. 8.12B. The linear glucose concentration range suits normal and diabetic people where the glucose level usually varies from 0.2 to 20 mM.

Similar to glucose, experiments were also performed with H_2O_2, and the relationship between the reduction current and the concentration of

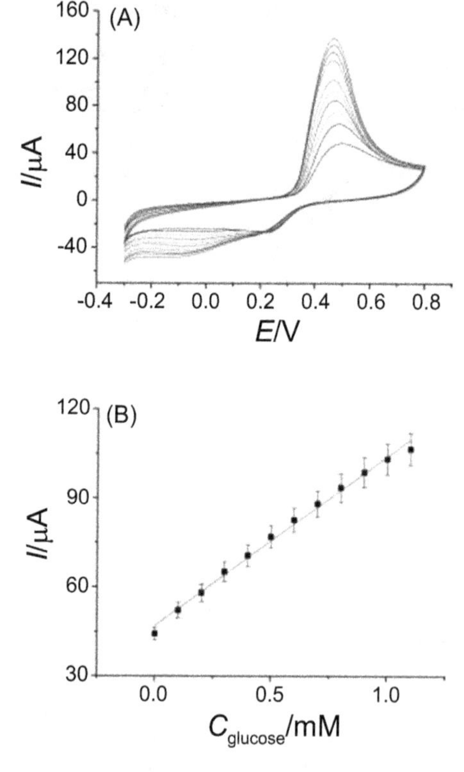

FIGURE 8.12 (A) CVs of CuS/CS/GCE in 0.10 M phosphate buffer solution (pH 7.2) containing glucose from 0 to 1.1 mM at the interval of 0.1 mM at 100 mV/s, and (B) the variation of oxidation peak current with the concentration of glucose. *CV*, Cyclic voltammetry; *GCE*, glassy carbon electrode. Source: *Figure reprinted from Y.J. Yang, J. Zi and W. Li, Enzyme-free sensing of hydrogen peroxide and glucose at a CuS nanoflowers modified glassy carbon electrode, Electrochim. Acta 115, 126–130 (2014) with permission from Elsevier.*

H_2O_2 was examined in a phosphate buffer solution (pH 7.2). The calibration curve for H_2O_2 is shown in Fig. 8.13 and it may be noted that the variation of current with the concentration of H_2O_2 from 1×10^{-6} to 1×10^{-4} M is linear. The sensitivity of the sensor for H_2O_2 concentration is 3.64×10^4 μA/M. One of the advantages of this study is that no enzyme has been used for sensing [51].

In another study on the detection of hydrogen peroxide using an electrochemical method, self-assembly of layered molybdenum disulfide-graphene (MoS_2–Gr) and horseradish peroxidase (HRP) was used [52]. To fabricate the probe, first the MoS_2–Gr composite was prepared by the chemical synthesis method and then it was dispersed in distilled water for 1 h and mixed with HRP. The HRP–MoS_2–Gr dispersion was added dropwise onto the GCE and then dried at 4°C in a

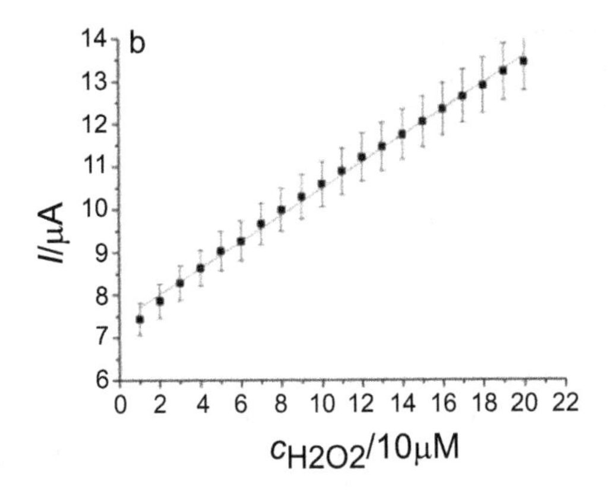

FIGURE 8.13 Variation of dynamic current response of CuS/CS/GCE with the concentration of H_2O_2 at applied potential −0.1 V. *GCE*, Glassy carbon electrode. Source: *Figure reprinted from Y.J. Yang, J. Zi and W. Li, Enzyme-free sensing of hydrogen peroxide and glucose at a CuS nanoflowers modified glassy carbon electrode, Electrochim. Acta 115, 126–130 (2014) with permission from Elsevier.*

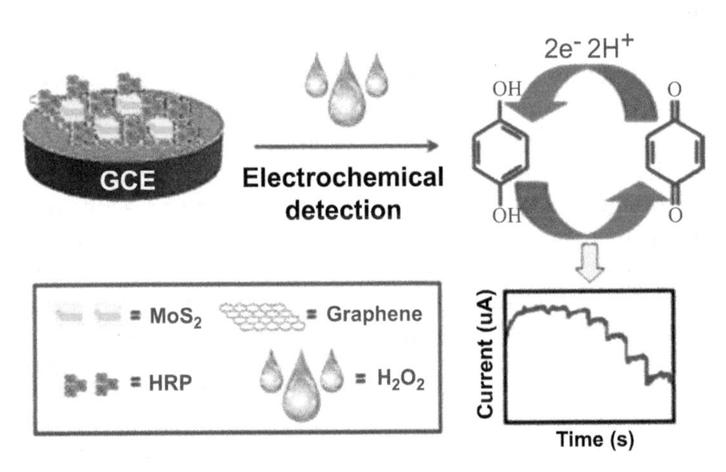

FIGURE 8.14 HRP–MoS_2–Gr GCE biosensor for amperometric analysis of H_2O_2. Source: *Figure reprinted from H. Song, Y. Ni and S. Kokot, Investigations of an electrochemical platform based on the layered MoS_2 - graphene and horseradish peroxidase nanocomposite for direct electrochemistry and electrocatalysis, Biosens. Bioelectron. 56, 137–143 (2014) with permission from Elsevier.*

refrigerator. The schematic of the H_2O_2 sensor using HRP–MoS_2–Gr composite is shown in Fig. 8.14. The chitosan solution was deposited on the modified electrode and dried in air to improve the stability of the modified electrode. The use of HRP–MoS_2–Gr nanocomposite biosensor for amperometric analysis of H_2O_2 is illustrated in Fig. 8.14. The HRP–MoS_2–Gr composite-based H_2O_2 sensor displays high sensitivity (679.7 μA/mM/cm^2), wide linear range (0.2 μM–1.103 mM), low detection limit (0.049 μM), and fast amperometric response.

Several glucose sensors have used metal chalcogenide where enzyme has not been used for the sensing mechanism. We have discussed one above and now we shall, briefly, mention a few of these. A non-enzymatic glucose sensor reported has used CuS/RGO/CuS nanocomposite [53]. The nanocomposite was grown on conductive Cu-foam through a hydrothermal-assisted process. In this sensor, CuS/RGO/CuS/Cu composite was directly used as the electrode and hence electrochemical sensor of glucose, and its performance was evaluated by cyclic voltammetry and amperometry techniques. A non-enzymatic glucose sensor in which Cu nanoparticles were decorated over MoS_2 nanosheet has also been reported [54]. For the fabrication, the suspension of Cu-MoS_2 prepared was dropped over the surface of bare GCE, and after this nafion was dropped over it. The so fabricated electrode was named GCE/Cu-MoS_2/Nafion. The samples of glucose were prepared in alkaline media and the performance of the sensor was evaluated by cyclic voltammetry and amperometric measurements. Electrochemical tests show a linear range of up to 4 mM of glucose concentration.

In a similar study utilizing the electrochemical method, AuNPs were decorated over MoS_2 nanosheets for the sensing of dopamine (DA) [55]. The electrode was modified with AuNPs@MoS_2 nanocomposite prepared by the microwave-assisted method. The sensor utilizes the combined advantages of MoS_2 and AuNPs. Further, the presence of AuNPs significantly increases the conductivity of the hybrid nanostructures. The nanocomposite film facilitates the electron transfer between DA present in the sample and the electrode and due to this, the modified electrode shows excellent sensitivity for DA. The oxidation peak current varies linearly with DA concentration in the range from 0.1 to 200 μM. It has been reported that the measurement of DA in the sample is not affected by the presence of large excess of ascorbic acid in the sample.

An electrochemical sensor, where the addition of MoS_2 wrinkles the graphene which further increases its surface area has been reported for the detection of eugenol [56]. For the fabrication of the probe, poly (diallyl dimethylammonium chloride) functionalized graphene-MoS_2 nanoflowers (PDDA-G-MoS_2) were first synthesized. After this, AuNPs were assembled on the surface of PDDA-G-MoS_2. The GCE was then modified using Au/PDDA-G-MoS_2 nanomaterial abbreviated as Au/PDDA-G-MoS_2/GCE. The modified electrode was then used to detect eugenol by electrochemical method. The oxidation peak current was found to increase distinctly by cyclic voltammetry (CV) in 0.10 mol/L NaAc-HAc buffer solution (pH = 5.50). The oxidation peak current of eugenol increases linearly with the addition of concentration in the range from 0.1 to 440 mol/L. This sensor has been reported to possess high sensitivity and good stability and can be used to determine eugenol concentration in real samples.

8.6.4 Colorimetric sensor

We have presented above the optical, electrical, and electrochemical biosensors utilizing metal chalcogenides. Now we discuss, briefly, a metal chalcogenide biosensor based on a colorimetric method to detect glucose in blood samples. It is based on the intrinsic peroxidase-like activity of tungsten disulfide (WS_2) nanosheets which catalyze the peroxidase substrate 3,3′,5,5′-tetramethylbenzidine (TMB) to produce a color reaction in the presence of H_2O_2 [57]. A portable test kit for the visual detection of blood glucose was developed using glucose oxidase (GOx) and WS_2 nanosheet-catalyzed reactions. The sensor possesses a linear range of operation for the glucose concentration from 5 to 300 μM. The portable test kit fabricated was used to determine the glucose levels in serum samples of normal and diabetic persons by observing the color change.

8.7 Summary

This chapter focuses on the fabrication of biosensors utilizing metal chalcogenides. For the best performance of the sensor, the material to be chosen for the fabrication is very important. Therefore, we have first discussed the selection of material and then the methods for its coating on the substrate to fabricate the sensor. The characterization techniques for the coated film/layer have been discussed. In the end, some of the recently reported metal chalcogenides-based biosensors utilizing optical, electrical, and electrochemical methods have been briefly described.

References

[1] N. Bhalla, P. Jolly, N. Formisano, P. Estrela, Introduction to biosensors, Essays Biochem. 60 (2016) 1–8.
[2] P. Damborsky, J. Svitel, J. Katrlik, Optical biosensors, Essays Biochem. 60 (2016) 91–100.
[3] M.A. Cooper, Optical biosensors in drug discovery, Nat. Rev. Drug. Discov. 1 (2002) 515–528.
[4] C. Li, Y. Wang, H. Jiang, X. Wang, Biosensors based on advanced sulfur-containing nanomaterials, Sensors 20 (2020) 3488.

[5] D. Chauhan, Pooja, V. Nirbhaya, C.M. Srivastava, et al., Nanostructured transition metal chalcogenide embedded on reduced graphene oxide based highly efficient biosensor for cardiovascular disease detection, Microchem. J 155 (2020) 104697.

[6] S. Wu, H. Huang, M. Shang, C. Du, Y. Wu, W. Song, High visible light sensitive MoS_2 ultrathin nanosheets for photoelectrochemical biosensing, Biosens. Bioelectron. 92 (2017) 646–653.

[7] E. Rahmanian, C.C. Mayorga-Martinez, R. Malekfar, J. Luxa, Z. Sofer, M. Pumera, 1T-Phase tungsten chalcogenides (WS_2, WSe_2, WTe_2) decorated with TiO_2 nanoplatelets with enhanced electron transfer activity for biosensing applications, ACS Appl. Nano Mater. 1 (2018) 7006–7015.

[8] Y. Hu, Y. Huang, C. Tan, X. Zhang, Q. Lu, M. Sindoro, et al., Two-dimensional transition metal dichalcogenide nanomaterials for biosensing applications, Mater. Chem. Front. 1 (2017) 24–36.

[9] S. Singh, P.K. Singh, A. Umar, P. Lohia, H. Albargi, L. Castaneda, et al., 2D nanomaterial-based surface plasmon resonance sensors for biosensing applications, Micromachines 11 (2020) 779.

[10] J. Lu, X. Zhang, N. Liu, X. Zhang, Z. Yu, T. Duan, Electrochemical detection of Cu^{2+} using graphene-SnS nanocomposite modified electrode, J. Electroanal. Chem. 769 (2016) 21–27.

[11] K. Wang, Q. Liu, Q.M. Guan, J. Wu, H.N. Li, J.J. Yan, Enhanced direct electrochemistry of glucose oxidase and biosensing for glucose via synergy effect of graphene and CdS nanocrystals, Biosens. Bioelectron. 26 (2011) 2252–2257.

[12] G. Suganthi, T. Arockiadoss, T.S. Uma, ZnS nanoparticles decorated graphene nanoplatelets as immobilization matrix for glucose biosensor, Nanosyst. Phys. Chem. Math. 7 (2016) 637–642.

[13] K.J. Huang, L. Wang, J. Li, Y.M. Liu, Electrochemical sensing based on layered MoS_2-graphene composites, Sens. Actuators B 178 (2013) 671–677.

[14] A.S. Hassanien, A.A. Akl, Effect of Se addition on optical and electrical properties of chalcogenide CdSSe thin films, Superlattices Microstruct. 89 (2016) 153–169.

[15] A. Hannachi, A. Segura, H.M. Meherzib, Growth of manganese sulfide (α-MnS) thin films by thermal vacuum evaporation: structural, morphological and optical properties, Mater. Chem. Phys. 181 (2016) 326–332.

[16] G. Wang, Q. Nie, X. Shen, F. Chen, J. Li, W. Zhang, et al., Phase change and optical band gap behavior of Ge–Te–Ga thin films prepared by thermal evaporation, Vacuum 86 (2012) 1572–1575.

[17] K.L. Choy, Chemical vapour deposition of coatings, Prog. Mater. Sci. 48 (2003) 57–170.

[18] J. Zhou, J. Lin, X. Huang, et al., A library of atomically thin metal chalcogenides, Nature 556 (2018) 355–359.

[19] J. You, M.D. Hossain, Z. Luo, Synthesis of 2D transition metal dichalcogenides by chemical vapor deposition with controlled layer number and morphology, Nano Converg. 5 (2018) 26.

[20] H.Y. Li, J.W. Yoon, C.S. Lee, K. Lim, J.W. Yoon, J.H. Lee, Visible light assisted NO_2 sensing at room temperature by CdS nanoflake array, Sens. Actuators B 255 (2018) 2963–2970.

[21] K.K. Liu, W. Zhang, Y.H. Lee, Y.C. Lin, M.T. Chang, C.Y. Su, et al., Growth of large-area and highly crystalline MoS_2 thin layers on insulating substrates, Nano Lett. 12 (3) (2012) 1538–1544.

[22] X. Li, X. Li, Z. Li, J. Wang, J. Zhang, WS_2 nanoflakes based selective ammonia sensors at room temperature, Sens. Actuators B 240 (2017) 273–277.

[23] N. Nakayama, K. Ito, Sprayed films of stannite Cu_2ZnSnS_4, Appl. Surf. Sci. 92 (1996) 171–175.

[24] M.A. Hossain, Z. Tianliang, L.K. Keat, L. Xianglin, R.R. Prabhakar, S.K. Batabyal, et al., Synthesis of Cu(InGa)(SSe)$_2$ thin films using an aqueous spray-pyrolysis approach, and their solar cell efficiency of 10.5%, J. Mater. Chem. A 3 (2015) 4147–4154.

[25] A. Nakaruk, C.C. Sorrell, Conceptual model for spray pyrolysis mechanism: fabrication and annealing of titania thin films, J. Coat. Technol. Res. 7 (2010) 665–676.

[26] C.N.R. Rao, H.S.S. Ramakrishna Matte, U. Maitra, Graphene analogues of inorganic layered materials, Angew. Chem. Int. Ed. 52 (2013) 13162–13185.

[27] X. Huang, Z. Zeng, H. Zhang, Metal dichalcogenide nanosheets: preparation, properties and applications, Che. Soc. Rev. 42 (2013) 1934–1946.

[28] M. Joshi, A. Bhattacharya, S.W. Ali, Characterization techniques for nanotechnology applications in textiles, Indian. J. Fiber Text. Res. 33 (2008) 304–317.

[29] R. Shrivastav, Synthesis and characterization techniques of nanomaterials, Int. J. Green. Nanotechnol. 4 (2012) 17–27.

[30] S. Boddolla, S. Thodeti, A review on characterization techniques of nanomaterials, Int. J. Eng. Sci. Math. 7 (2018) 169–175.

[31] B.D. Cullity, Elements of X-ray Diffraction, Addison-Wesley, USA, 1978.

[32] D. Yang, S.J. Sandoval, W.M.R. Divigalpitiya, J.C. Irwin, R.F. Frindt, Structure of single-molecular-layer MoS$_2$, Phys. Rev. B 43 (1991) 12053–12056.

[33] G. Singh, B. Rakesh, M. Sharma, Raman spectroscopy: basic principle, instrumentation and selected applications for the characterization of drugs of abuse, Egypt. J. Forensic Sci. 6 (2016) 209–215.

[34] B.H. Stuart, Infrared Spectroscopy: Fundamentals and Applications, Wiley, USA, 2004.

[35] B.D. Gupta, R. Kant, Recent advances in surface plasmon resonance based fiber optic chemical and biosensors utilizing bulk and nanostructures, Opt. Laser Technol. 101 (2018) 144–161.

[36] S.P. Usha, A.M. Shrivastav, B.D. Gupta, Semiconductor metal oxide/polymer based fiber optic lossy mode resonance sensors: a contemporary study, Opt. Fiber Technol. 45 (2018) 146–166.

[37] S. Kaushik, U.K. Tiwari, A. Deep, R.K. Sinha, Two-dimensional transition metal dichalcogenides assisted biofunctionalized optical fiber SPR biosensor for efficient and rapid detection of bovine serum albumin, Sci. Rep. 9 (2019) 6987.

[38] B. Du, Y. Ruan, T.T. Ly, P. Jia, Q. Sun, Q. Feng, et al., MoS$_2$-enhanced epoxy-based plasmonic fiber-optic sensor for selective and sensitive detection of methanol, Sens. Actuat. B 305 (2020) 127513.

[39] Q. Ouyang, S. Zeng, L. Jiang, L. Hong, G. Xu, X.Q. Dinh, et al., Sensitivity enhancement of transition metal dichalcogenides/silicon nanostructure-based surface plasmon resonance biosensor, Sci. Rep. 6 (2016) 28190.

[40] S.P. Usha, B.D. Gupta, Urinary p-cresol diagnosis using nanocomposite of ZnO/MoS$_2$ and molecular imprinted polymer on optical fiber based lossy mode resonance sensor, Biosens. Bioelectron. 101 (2018) 135–145.

[41] S. Sharma, A.M. Shrivastav, B.D. Gupta, Lossy mode resonance based fiber optic creatinine sensor fabricated using molecular imprinting over nanocomposite of MoS$_2$/SnO$_2$, IEEE Sens. J. 20 (2020) 4251–4259.

[42] K.H. Han, J.Y. Kim, S.G. Jo, C. Seo, J. Kim, J. Joo, Sensitive optical bio-sensing of p-type WSe$_2$ hybridized with fluorescent dye attached DNA by doping and de-doping effects, Nanotechnology 28 (2017) 435501.

[43] Q. Xi, D.M. Zhou, Y.Y. Kan, J. Ge, Z.K. Wu, R.Q. Yu, et al., Highly sensitive and selective strategy for microrna detection based on WS$_2$ nanosheet mediated fluorescence quenching and duplex-specific nuclease signal amplification, Anal. Chem. 86 (2014) 1361–1365.

[44] D. Sarkar, W. Liu, X. Xie, A.C. Anselmo, S. Mitragotri, K. Banerjee, MoS_2 field-effect transistor for next-generation label-free biosensors, ACS Nano 8 (2014) 3992–4003.
[45] K. Lee, R. Gatensby, N. McEvoy, T. Hallam, G.S. Duesberg, High-performance sensors based on molybdenum disulfide thin films, Adv. Mater. 25 (2013) 6699–6702.
[46] D. Sarkar, X. Xie, J. Kang, H. Zhang, W. Liu, J. Navarrete, et al., Functionalization of transition metal dichalcogenides with metallic nanoparticles: implications for doping and gas-sensing, Nano Lett. 15 (2015) 2852–2862.
[47] L. Wang, Y. Wang, J.I. Wong, T. Palacios, J. Kong, H.Y. Yang, Functionalized MoS_2 nanosheet-based field-effect biosensor for label-free sensitive detection of cancer marker proteins in solution, Small 10 (2014) 1101–1105.
[48] X. Liu, H.L. Shuai, Y.J. Liu, K.J. Huang, An electrochemical biosensor for DNA detection based on tungsten disulfide/multi-walled carbon nanotube composites and hybridization chain reaction amplification, Sens. Actuators B 235 (2016) 603–613.
[49] K.J. Huang, Y.J. Liu, H.B. Wang, et al., Sub-femtomolar DNA detection based on layered molybdenum disulfide/multi-walled carbon nanotube composites, Au nanoparticle and enzyme multiple signal amplification, Biosens. Bioelectron. 55 (2014) 195–202.
[50] A.H. Loo, A. Bonanni, A. Ambrosi, M. Pumera, Molybdenum disulfide (MoS_2) nanoflakes as inherently electroactive labels for DNA hybridization detection, Nanoscale 6 (2014) 11971–11975.
[51] Y.J. Yang, J. Zi, W. Li, Enzyme-free sensing of hydrogen peroxide and glucose at a CuS nanoflowers modified glassy carbon electrode, Electrochim. Acta 115 (2014) 126–130.
[52] H. Song, Y. Ni, S. Kokot, Investigations of an electrochemical platform based on the layered MoS_2 - graphene and horseradish peroxidase nanocomposite for direct electrochemistry and electrocatalysis, Biosens. Bioelectron. 56 (2014) 137–143.
[53] C. Zhao, X. Wu, X. Zhang, P. Li, X. Qian, Facile synthesis of layered CuS/RGO/CuS nanocomposite on Cu foam for ultrasensitive nonenzymatic detection of glucose, J. Electroanal. Chem. 785 (2017) 172–179.
[54] J. Huang, Z. Dong, Y. Li, J. Li, W. Tang, H. Yang, et al., MoS_2 nanosheet functionalized with Cu nanoparticles and its application for glucose detection, Mater. Res. Bull. 48 (2013) 4544–4547.
[55] S. Su, H. Sun, F. Xu, L. Yuwen, L. Wang, Highly sensitive and selective determination of dopamine in the presence of ascorbic acid using gold nanoparticles-decorated MoS_2 nanosheets modified electrode, Electroanalysis 25 (2013) 2523–2529.
[56] Q. Feng, K. Duan, X. Ye, D. Lu, Y. Du, C. Wang, A novel way for detection of eugenol via poly(diallyldimethylammonium chloride) functionalized graphene-MoS_2 nano-flower fabricated electrochemical sensor, Sens. Actuators B 192 (2014) 1–8.
[57] T. Lin, L. Zhong, Z. Song, L. Guo, H. Wu, Q. Guo, et al., Visual detection of blood glucose based on peroxidase-like activity of WS_2 nanosheets, Biosens. Bioelectron. 62 (2014) 302–307.

CHAPTER

9

Selectivity, sensitivity, and time factors

9.1 Introduction

In Chapter 8, we have discussed the fabrication and characterization of sensor probes utilizing metal chalcogenide. The next step after the fabrication of the sensor is the evaluation of its performance which is the important part of the sensor. Its usage in the field depends on the performance parameters. The performance of the fabricated sensors is evaluated in terms of performance parameters like sensitivity, the limit of detection (LOD), the figure of merit (FOM), selectivity, stability, repeatability, response time, recovery time, and lifetime. Further, knowledge of these parameters is required to compare various sensors. The definitions of some of these parameters depend on the method of interrogation used for sensing. These parameters also set the limitations and operating and environmental conditions for the user before its use.

As mentioned above, the interrogation method used for sensing is also important for the evaluation of the performance of the sensor. For example, in the case of SPR-based sensors, angular, wavelength, and intensity interrogations are used to calibrate the sensor. Accordingly, the resolution of the detection system such as the photodetector, and the resolving power of the grating used in the spectrometer affect the sensitivity and the limit of detection of the sensor. To evaluate the sensor, for a particular application, the performance parameters are figured out from the sensor response curve. In the case of optical sensors, the response can be the variation of transmittance power, reflectivity, absorbance, or fluorescence intensity as a function of analyte concentration. In this chapter, we shall discuss various performance parameters of the

Metal Chalcogenide Biosensors
DOI: https://doi.org/10.1016/B978-0-323-85381-1.00002-7

optical, electrochemical, and electrical-based biosensors fabricated using metal chalcogenides by taking examples from various studies reported in the literature.

9.2 Performance evaluation parameters

The performance of a biosensor is evaluated in terms of many parameters. Below we discuss them one by one using examples of various sensors.

9.2.1 Sensitivity

Among the various performance parameters of biosensors, sensitivity is one of the most important parameters. Apart from sensing capability, it decides how low/minimum change in the value of measurand/analyte can be detected/sensed by the sensor. It is, generally, defined as the quantitative change in the sensor's response signal for a measurable unit change in the quantity/property of the measurand/analyte. The response signal depends on the approach that has been decided before the fabrication of the biosensor. For example, the sensor may be based on optical, electrochemical, thermal, or electrical methods. In the case of sensors based on the optical method, the response can be in terms of a change in the wavelength, phase, intensity, or polarization of the input light source. In electrochemical sensors, the variation may occur in current, potential, resistance, or capacitance. Thus, the sensor response may change according to the type of the sensor and therefore the definition of the sensitivity of the sensor will change accordingly. The sensitivity of the sensor is defined by the ratio of the change in sensor response output to the change in the measurand or, in other words, it is the derivative of the calibration curve that gives the sensitivity of the sensor. The calibration curve is the plot between the sensor response and the measurand. The definition of sensitivity in the case of electrochemical sensors changes due to the dependence of response on the area of the electrode. It is defined as the slope of the calibration curve divided by the area of the electrode. For plotting the calibration curve, first, the response is recorded for different values of measurand. The calibration curve can have any mathematical variation but the sensor with linear variation is considered to be the best performing sensor since it provides its maximum sensitivity for the complete range of operation. In the last chapter, Section 8.6.1, we discussed one SPR-based BSA sensor utilizing the MoS_2 layer [1]. The SPR spectra for different concentrations of BSA sample using a probe with MoS_2 layer have been plotted in

Fig. 8.3C. These plots are the responses for different concentrations of BSA which is the measurand. From these spectra, resonance wavelengths were determined for different BSA concentrations and the data so obtained were used to plot the calibration curve of the sensor shown in Fig. 8.3D. In this case, the calibration curve is linear. The sensitivity of the sensor, in this case, is defined as the change in the resonance wavelength per unit change in the concentration of BSA which is the slope of this curve, and its value so determined is 0.9234 nm/(μg/mL) [1]. As mentioned, in the case of linear variation of the output signal with the input measurand, the sensitivity remains constant for the entire range of operation of the sensor as obtained in this case. However, it does not remain constant in all cases of biosensors.

Another example where the response is linear is the detection of copper ion (Cu^{2+}) using an electrochemical sensor based on a glass carbon electrode (GCE) modified with SnS nanosheets supported on graphene (Gr–SnS) [2]. The effect of Cu^{2+} on electrochemical response was studied using cyclic voltammetry and linear sweep voltammetry (LSV). The LSV curves of Gr-SNS/GCE for different concentrations of Cu^{2+} in 1.0 M KCl solution of pH 4.5 are shown in Fig. 9.1A. The concentration of Cu^{2+} in the figure is increasing from a to i. The peak current decreases with increasing concentration of Cu^{2+} in the range from 1.5 to 36.0 μM and has been shown in Fig. 9.1B. In this case, the variation is linear. The sensitivity determined from the slope of the curve is 0.241 μA/μM.

An H_2O_2 biosensor has been reported based on the dependence of the catalytic activity of WS_2 nanosheets on H_2O_2 [3]. It is a visual method where the addition of H_2O_2 in a mixture of WS_2 nanosheets, 3,3′,5,5′-tetramethylbenzidine (TMB), and Tris–HCl buffer changes the color of the mixture. For quantitative measurements, the absorbance of the mixture was recorded at 450 nm wavelength. Fig. 9.2A shows the variation of absorbance with a concentration of H_2O_2 in the range of 0 to 300 μM. In this case, the calibration curve is non-linear. However, if it is plotted for a short range (10 to 100 μM) then it can be taken as linear as shown in Fig. 9.2B. The sensitivity of the sensor for this particular range can be determined by taking the slope of the linear curve.

Now we discuss a non-enzymatic glucose sensor where the calibration curve is non-linear and sensitivity has been reported for two ranges of glucose concentration. The sensor was fabricated by in-situ growing layered CuS/RGO/CuS nanostructure on the surface of Cu foam [4]. The CuS/RGO/CuS/Cu was directly used as an electrochemical sensor of glucose. Its performance was evaluated using cyclic voltammetry and amperometry techniques. The amperometric response of the electrode was studied by successive addition of glucose in 0.1 M NaOH solution in optimized conditions. Fig. 9.3A shows the enhanced response current

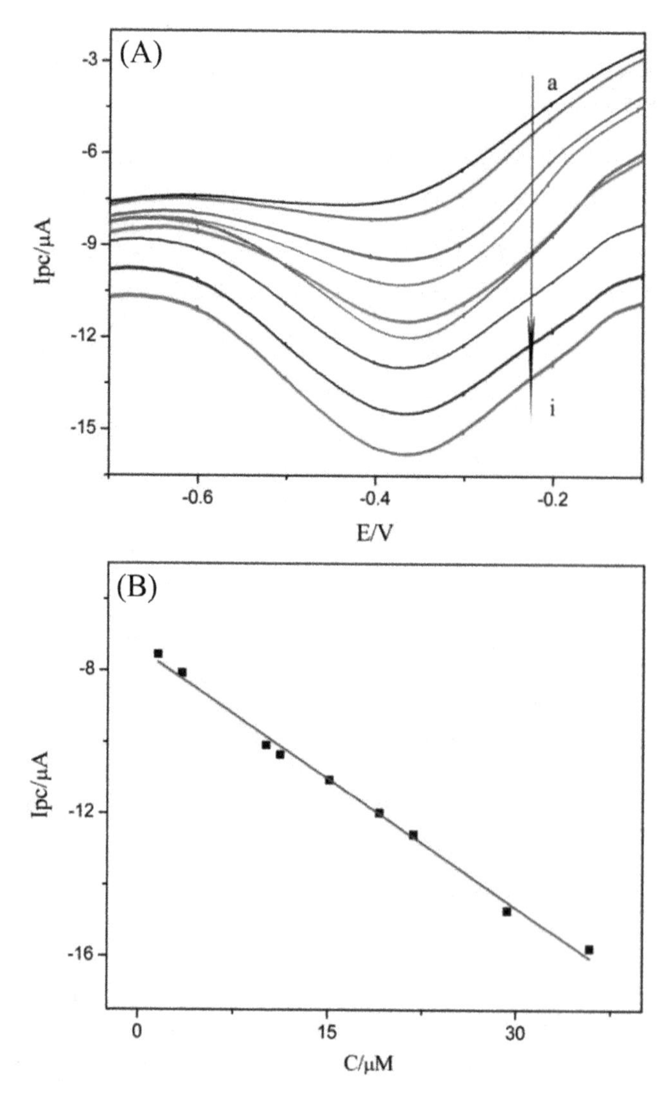

FIGURE 9.1 (A) LSV curves of the Gr–SnS/GCE for different concentrations of Cu^{2+}, and (B) variation of peak current with a concentration of Cu^{2+}. *GCE*, Glass carbon electrode; *LSV*, linear sweep voltammetry. Source: *Figure reprinted from J. Lu, X. Zhang, N. Liu, X. Zhang, Z. Yu, T. Duan, Electrochemical detection of Cu^{2+} using graphene–SnS nanocomposite modified electrode, J. Electroanalyt. Chem. 769, 21–27 (2016) with permission from Elsevier.*

on successive addition of glucose. The calibration curve of the sensor is shown in Fig. 9.3B. Its variation is non-linear but has been divided into two linear ranges. A lower concentration range from 1 to 655 μM with a sensitivity of 22.67 mA/(mM.cm^2) and a higher concentration range

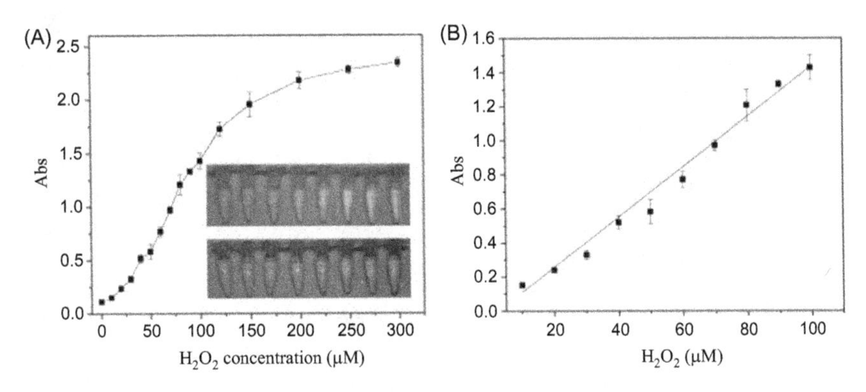

FIGURE 9.2 (A) Response curve of WS_2-based H_2O_2 sensor, and (B) linear calibration plot for a short range of H_2O_2 concentration. Source: *Figure reprinted from T. Lin, L. Zhong, Z. Song, L. Guo, H. Wu, Q. Guo, et al., Visual detection of blood glucose based on peroxidase-like activity of WS_2 nanosheets, Biosens. Bioelectron. 62, 302–307 (2014) with permission from Elsevier.*

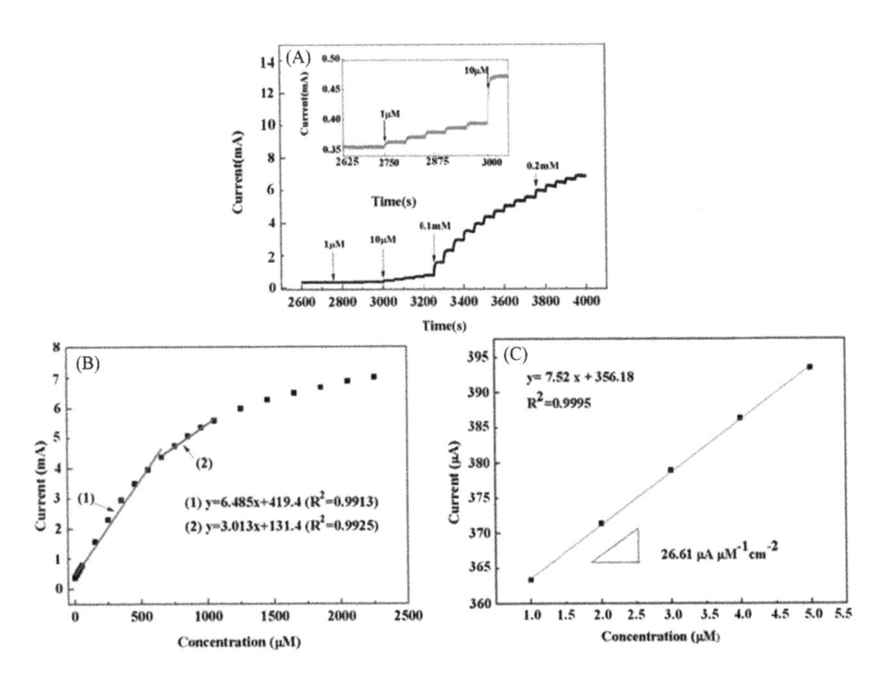

FIGURE 9.3 (A) Amperometric response, (B) the calibration curve of the CuS/RGO/CuS/Cu foam electrode with successive addition of glucose in 0.1 M NaOH, and (C) the calibration curve for the successive addition of 1 μM glucose. Source: *Figure reprinted from C. Zhao, X. Wu, X. Zhang, P. Li, X. Qian, Facile synthesis of layered CuS/RGO/CuS nanocomposite on Cu foam for ultrasensitive nonenzymatic detection of glucose, J. Electroanalyt. Chem. 785, 172–179 (2017) with permission from Elsevier.*

starting from 0.655 to 1.055 mM with a lower sensitivity of 10.54 mA/(mM.cm^2). For a very low concentration range, the sensitivity determined is 26.61 mA/(mM.cm^2). To calculate the sensitivity, the area of the electrode used was 0.283 cm^2.

The last example for the sensitivity determination is of glucose sensor utilizing a MoS_2 nanosheet functionalized with Cu nanoparticles (Cu-MoS_2 hybrid) over GCE [5]. A coating of Nafion was also performed over the hybrid. It is a non-enzymatic sensor and its performance was evaluated using cyclic voltammetry and amperometric measurements in alkaline media. Similar to the previous example, the response of GCE/Cu-MoS_2/Nafion was recorded using successive addition of glucose into 0.1 M NaOH solution at an optimal potential of 0.65 V. Fig. 9.4A shows the enhancement of response on successive addition of glucose. The current increases as the concentration of glucose increases as shown in Fig. 9.4B. For some parts (0 to 4 mM), the variation was taken as linear. The sensitivity calculated using the slope of the calibration curve is 1055 μM/(mM.cm^2). The electrode diameter was 3 mm.

9.2.2 Limit of detection

The limit of detection (LOD) is the lowest concentration of the target molecule in a solution that can be detected by the sensor. Although its definition and concept are simple, there is always an uncertainty in the calculation of its value depending on the technique, calibration, and the noises/errors involved in the measurements. Different methods have been used for the determination of LOD but the most common method for evaluating the LOD of a sensor is based on the signal-to-noise ratio of the device. In this method, LOD is calculated by taking a multiplication factor of 3 with the standard deviation near the blank concentration. Mathematically, it is written as

$$\mathrm{LOD} = 3\left(\frac{\sigma}{S_o}\right)$$

where σ is the standard deviation calculated in the sensor response while S_0 denotes the slope of the calibration curve or the sensitivity of the sensor near zero concentration of the analyte. The standard deviation in the response near zero concentration is determined by repeating the experiment a number of times at the lowest concentration. The LODs determined using the above formula for the examples discussed above for sensitivity are 0.02 [2], 1.2 [3], and 0.5 μM [4]. It should be remembered that LOD is a very crucial parameter for the viability of the sensor in real applications.

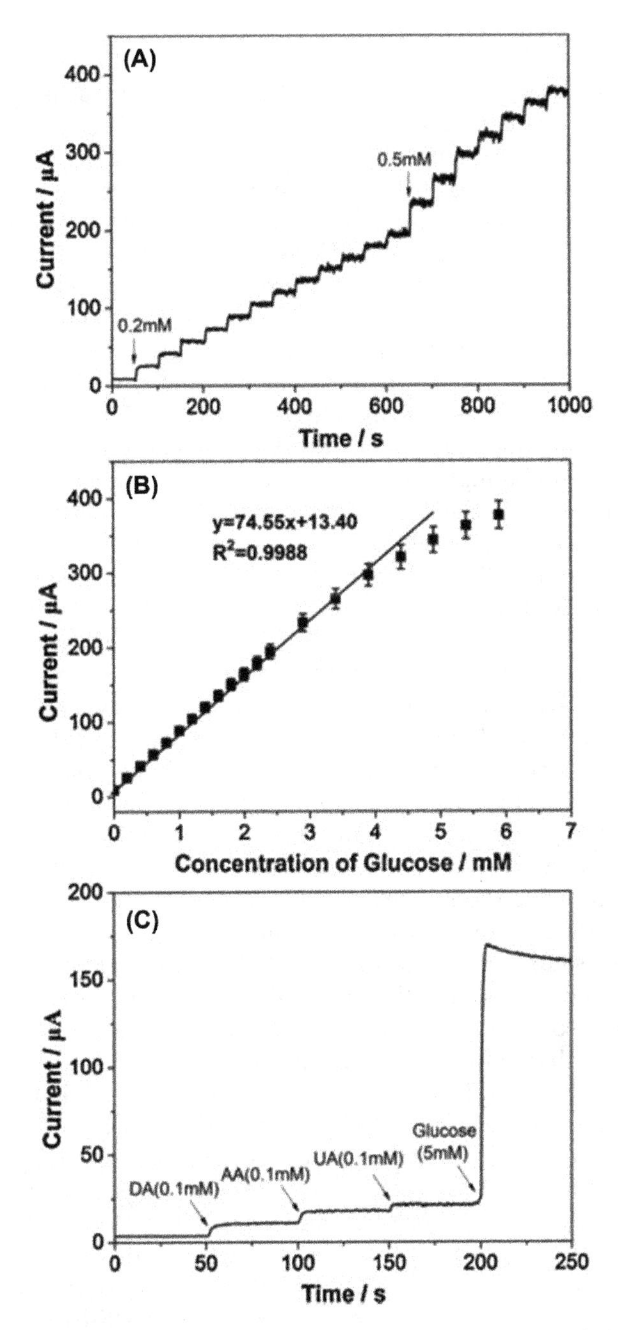

FIGURE 9.4 (A) Amperometric responses with successive addition of glucose, (B) the calibration curve, and (C) amperometric response to interferences from DA, AA, and UA at physiological concentration for GCE/Cu-MoS_2/Nafion. *Source: Figure reprinted from J. Huang, Z. Dong, Y. Li, J. Li, W. Tang, H. Yang, et al., MoS_2 nanosheet functionalized with Cu nanoparticles and its application for glucose detection, Mater. Res. Bull. 48, 4544–4547 (2013) with permission from Elsevier.*

9.2.3 Selectivity

Selectivity is another most important parameter of a sensor as it governs its utility for complex environments where other interferants are also present along with the analyte of interest. For ensuring selectivity, the design of the sensing layer over the probe is a very pivotal step for the sensor. The sensing layer ensures a specific interaction mechanism with the analyte of interest and rejects other interferants. In this way, the sensor shows the maximum response for the particular analyte and remains inactive for others. Hence, selectivity is a decisive performance parameter for a sensor to be applicable in real scenarios. However, it is rare to find a sensor that responds to only one analyte. Generally, a sensor responds mainly to one analyte with a limited response to other similar analytes. For good sensing characteristics, the selectivity of the sensor should be high, that is, a sensor should respond highly to the sensing analyte and for other analytes, its response should be low. The ratio or percentage of selectivity shown for a desired measurand/analyte by the sensor in the presence of other measurands is also classified under selectivity. The response of other measurands on the sensing of a particular analyte/measurand is called cross-selectivity. This exists when multiple physical/chemical/biological parameters can act as measurands in a particular analyte [6]. Sometimes, selectivity is also called specificity which is defined as the degree of accuracy to which the desired analyte can be sensed in a mixture or real sample [7]. The sensor extremely selective for a particular analyte is considered specific [8]. Selectivity is a parameter that can be graded or calibrated accordingly whereas specificity is a definite parameter [9]. However, the conditions and definition of rendering selectivity and sensitivity differ depending on the experimental and environmental conditions, analyte, sample/mixture/real sample in which the analyte has to be sensed, the method of analysis, and the quantitative or qualitative determination [7,8]. Now we shall give some examples of measurements for selectivity tests of a few biosensors based on metal chalcogenide.

Fig. 9.4(C) shows amperometric responses of GCE/Cu-MoS_2/Nafion to dopamine (DA), ascorbic acid (AA), uric acid (UA), and glucose for their physiological concentrations [5]. The current responses given by DA, AA, and UA are 4.8%, 5.2%, and 2.1% in comparison to glucose, respectively, which are negligible, and hence one can interpret that the sensor is selective to glucose. The presence of DA, AA, and UA in glucose samples as interferants cannot affect much the measurement of glucose concentration.

The selectivity test was also performed in an LMR-based creatinine sensor fabricated using molecular imprinting over a nanocomposite of MoS_2/SnO_2 [10]. The sensor has been discussed in Section 8.6.1. The sensor probe possesses four layers of the fiber core, LMR supporting layer

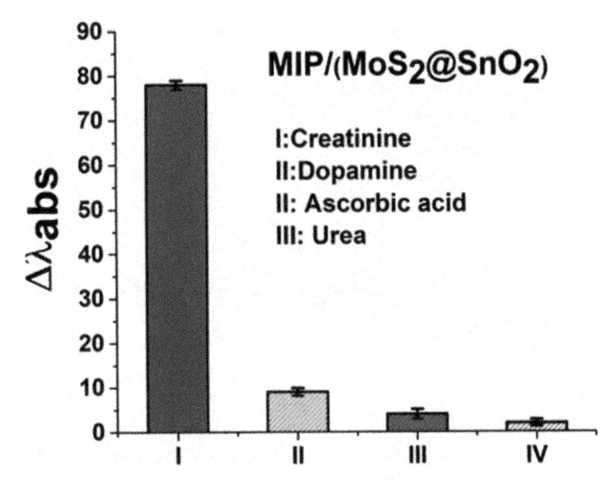

FIGURE 9.5 Shift in peak absorbance wavelength for the change in concentration from 0 to 2000 μg/mL of creatinine, DA, AA, and urea in the case of an LMR-based sensor. Source: *Figure reprinted from S. Sharma, A.M. Shrivastav, B.D. Gupta, Lossy mode resonance based fiber optic creatinine sensor fabricated using molecular imprinting over nanocomposite of MoS_2/SnO_2, IEEE Sens. J. 20, 4251–4259 (2020) with permission from IEEE.*

(MoS_2@SnO_2 nanocomposite), recognition layer [MIP/(MoS_2@SnO_2)], and sample solution. The absorbance spectra were measured for different concentrations of creatinine to calibrate the sensor. To test the selectivity, the shift in peak absorbance wavelength was measured for the change in concentration of creatinine, DA, AA, and urea from 0 to 2000 μg/mL. The shift obtained is shown in Fig. 9.5. The shifts in peak absorbance wavelength for DA, AA, and urea are negligible in comparison to creatinine which implies that the sensor is highly selective to creatinine.

An electrochemical sensor utilizing gold nanoparticles on the surface of poly (diallyl dimethylammonium chloride) functionalized graphene-MoS_2 nano-flower materials over glassy carbon electrode (Au/PDDA-G-MoS_2/GCE) has been reported for the sensing of eugenol [11]. The amperometric response (steady state response current) of the sensor to eugenol in the concentration range from 0.10 mol/L to 440 mol/L for the optimized conditions was found to increase linearly. Its selectivity was tested using common inorganic ions, some saccharines, and other organic compounds in the presence of 10 μmol/L eugenols. The effect of these interferants is negligible as shown in Fig. 9.6. Thus the sensor is highly selective to eugenol.

9.2.4 Limit of quantification

The limit of quantification (LOQ) is defined as the lowest analyte concentration that can be sensed and quantified by the sensor within

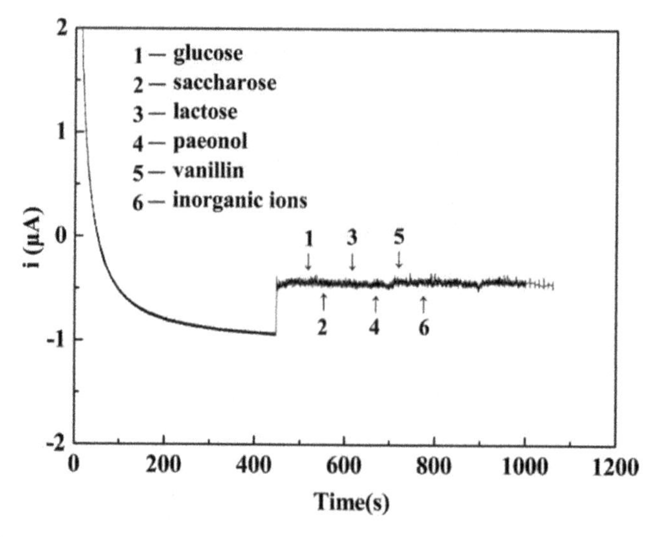

FIGURE 9.6 Interference of different species in the sensing of eugenol with the Au/PDDA-G-MoS_2/GCE-based electrochemical sensor. Source: *Figure reprinted from Q. Feng, K. Duan, X. Ye, D. Lu, Y. Du, C. Wang, A novel way for detection of eugenol via poly(diallyldimethylammonium chloride) functionalized graphene-MoS2 nano-flower fabricated electrochemical sensor, Sens. Actuat. B 192, 1–8 (2014) with permission from Elsevier.*

the limit of certainty after repeating experiments a number of times. Here it should be noticed that the LOD and LOQ of the sensor are rarely the same. Usually, the value of LOQ is higher than that of LOD [12]. Generally, the LOQ of a sensor is evaluated as 10 times the ratio of standard deviation in sensor response to the sensitivity near zero concentration of analyte and is expressed as [13]

$$\text{LOQ} = 10\left(\frac{\sigma}{S_o}\right)$$

where σ is the standard deviation in the sensor response and S_o denotes the slope of the calibration curve or the sensitivity of the sensor near zero concentration.

9.2.5 Figure of merit

This parameter is important for SPR-based biosensors. It relates to the sensitivity of the sensor and the detection accuracy of the measuring signal parameter (wavelength/angle). In such kinds of sensors, SPR spectra are recorded and from these spectra the resonance wavelengths/angles are determined. The resonance parameter corresponds to the dip in the SPR spectrum and hence its determination depends on the width of the SPR spectrum. The narrow spectrum gives a more accurate value of the resonance parameter. For the best performance of the sensor, the sensitivity should be high and the full width at half maximum (FWHM) of the SPR curve should be narrow. The figure of merit (FOM) of the sensor is defined as the ratio of the sensitivity (S) to FWHM. Mathematically, it is written as

$$\mathrm{FOM} = \frac{S}{\mathrm{FWHM}}$$

For better sensor performance, FOM should be high. To achieve its high value, the sensitivity should be high while FWHM should be small. Various combinations of metals have been reported to achieve a high value of FOM for SPR-based biosensors.

9.2.6 Repeatability

The repeatability test of a sensor is performed by repeating experiments using samples of two different concentrations several times and recording their responses. The variations in the measured parameter/response after performing the experiment repeatedly show whether the sensor satisfies the repeatability test or not. Mathematically, in the case of sensors, it is the changeability in the median value of response when the same experiment is repeated a number of times by the same person. It also depends on the accuracy to which the measurements are repeated. To give examples, we have chosen a few studies. The first study is the creatinine sensor based on the LMR phenomenon and discussed in Section 8.6.1. The repeatability test of the sensor was performed using samples of 100 and 2000 μg/mL concentrations of creatinine [10]. First, the LMR spectrum was recorded for 100 μg/mL and after 1 min it was again recorded. From these spectra, the peak absorbance wavelengths (λ_{abs}) were noted. The 100 μg/mL sample was then replaced by another sample of 2000 μg/mL concentration in the flow cell and its spectrum and λ_{abs} were recorded. The LMR spectrum and the value of λ_{abs} were again recorded after another 1 min. After this, the sample of 2000 μg/mL was replaced by a 100 μg/mL sample, and the corresponding λ_{abs} value was recorded. This completed one cycle of the repeatability experiment. The cycle was repeated to confirm the repeatable values of λ_{abs}. Fig. 9.7 shows the values of λ_{abs} for three cycles of the experiments performed using samples of 100 and 2000 μg/mL concentrations of creatinine and supports the repeatability of the response of the probe.

A similar procedure was adopted for the repeatability test of the p-cresol LMR-based sensor [14]. The sensor has been discussed in Section 8.6.1. First, the absorbance spectra were recorded without a sample and after 10 s, the p-cresol sample of 1000 μM concentration was

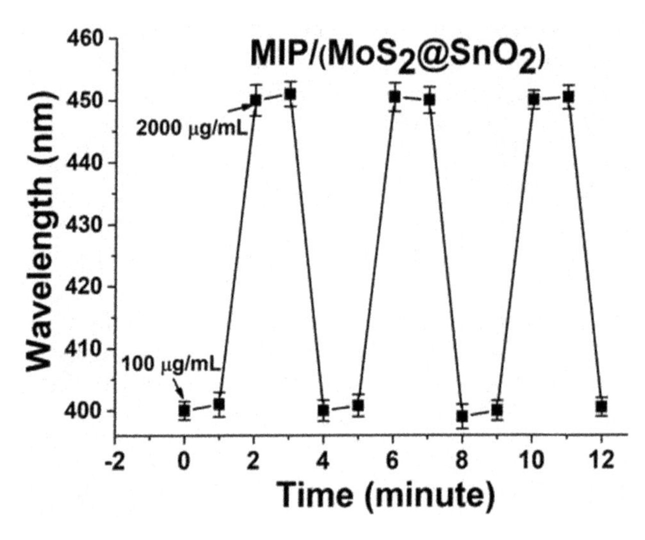

FIGURE 9.7 Repeatability test of LMR-based creatinine sensor. Source: *Figure reprinted from S. Sharma, A.M. Shrivastav, B.D. Gupta, Lossy mode resonance based fiber optic creatinine sensor fabricated using molecular imprinting over nanocomposite of* MoS_2/SnO_2*, IEEE Sens. J. 20, 4251–4259 (2020) with permission from IEEE.*

poured into the flow cell. Its absorbance spectra were recorded continuously. After this, the sample was removed and absorbance spectra were recorded continuously. This completed one cycle of measurements. The cycle was repeated to find the response. From the absorbance spectra absorbance values were determined for 462 nm wavelength. The results obtained for two cycles are shown in Fig. 9.8. The repeatable absorbance values confirm the repeatability of the sensor probe.

9.2.7 Detection accuracy

Detection accuracy is the experimental precision in the result that can be achieved by the sensor. The performance of the sensor and the results obtained from it are affected by environmental noises and the resolutions of the equipment used in the sensor design including human error. Thus, the percentage of uncertainty in measurement in comparison to the standard result is the detection accuracy. The deviation of the measured value from the standard value determines the accuracy.

9.2.8 Time factors

The time factors are also important in the case of biosensors. This is because the main part of the probe is the sensing layer. Its durability,

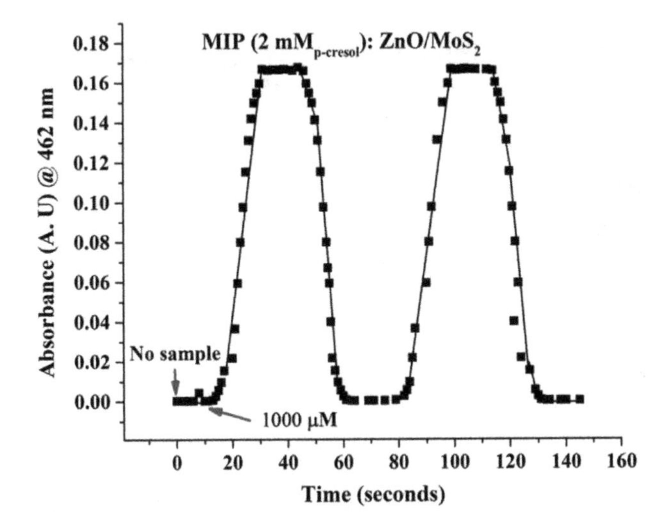

FIGURE 9.8 Repeatability test of LMR-based p-cresol sensor. Source: *Figure reprinted from S.P. Usha, B.D. Gupta, Urinary p-cresol diagnosis using a nanocomposite of ZnO/MoS$_2$ and molecular imprinted polymer on optical fiber-based lossy mode resonance sensor, Biosens. Bioelectron. 101, 135–145 (2018) with permission from Elsevier.*

reaction time with an analyte, and recovery after removal of analyte surrounding it have a certain definite time. A sensor cannot provide the same result for an indefinite period because the performance of the sensor deteriorates after a certain period. Apart from this the reaction between the analyte and the sensing layer to reach equilibrium takes time. Thus, below, we briefly present the time factors related to biosensors.

9.2.8.1 Response and recovery times

It is the time required for the sensor to give a stable output signal after bringing the sample in contact with the probe. When an analyte to be sensed is brought in contact with the biosensor, an interaction between the sensing layer and the analyte starts which takes a certain time to complete. During this interaction, the optical or electrical signal continues to change and saturates when the interaction is completed or equilibrium is reached. The time taken for saturation of the signal is called the response time of the biosensor. The response time in biosensors can be from a few seconds to a few minutes or even a few hours depending on the reaction. In the present scenario, a fast response is very important for the early detection of disease in clinical applications or early leakage warnings of gases for environmental applications. Further, continuous online monitoring of an analyte is possible only if the sensor probe responds quickly to the analyte. As SPR-based sensors

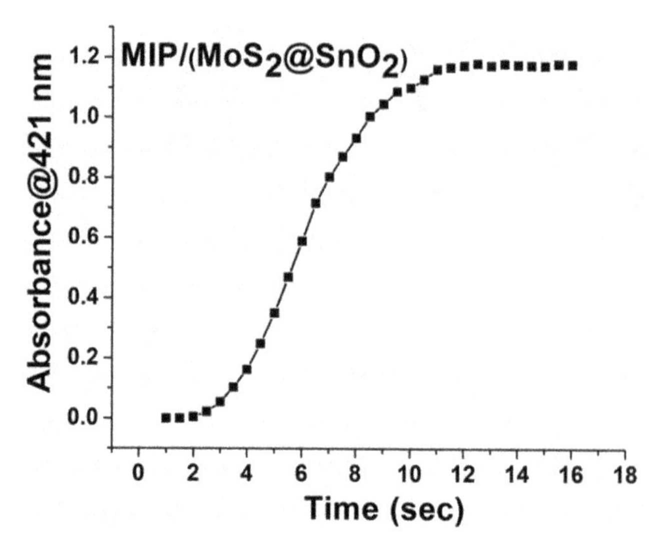

FIGURE 9.9 Response time of LMR-based creatinine sensor. Source: *Figure reprinted from S. Sharma, A.M. Shrivastav, B.D. Gupta, Lossy mode resonance based fiber optic creatinine sensor fabricated using molecular imprinting over nanocomposite of MoS_2/SnO_2, IEEE Sens. J. 20, 4251–4259 (2020) with permission from IEEE.*

provide real-time information on kinetic surface changes, they have a very fast response time. Here we give two examples of measurement of the response time of biosensors that have used metal chalcogenide in fabricated probes.

The first example is that of a creatinine sensor based on the LMR phenomenon and discussed in Section 8.6.1. The response time of this sensor was evaluated using a creatinine sample of 400 μg/mL concentration in the flow cell and measuring the variation of absorbance with time at 421 nm wavelength [10]. Fig. 9.9 shows the increase in absorbance with time after pouring the sample. It saturates after approximately 10 s which implies that the response time of the sensor is 10 s. In this sensor, the response is fast due to the non-covalent binding of creatinine molecules with binding sites of the molecularly imprinted polymer layer.

Another example is that of the p-cresol sensor which has also been discussed in Section 8.6.1. The measurement of response time, in this case, is similar to the previous example. It is evaluated by measuring the time variation of the absorbance at 462 nm wavelength after pouring the p-cresol sample of 1000 μM concentration into the flow cell [14]. The variation is shown in Fig. 9.10. The absorbance becomes constant after 15 s of the addition of a sample of 1000 μM. Thus, the response time of the sensor is 15 s.

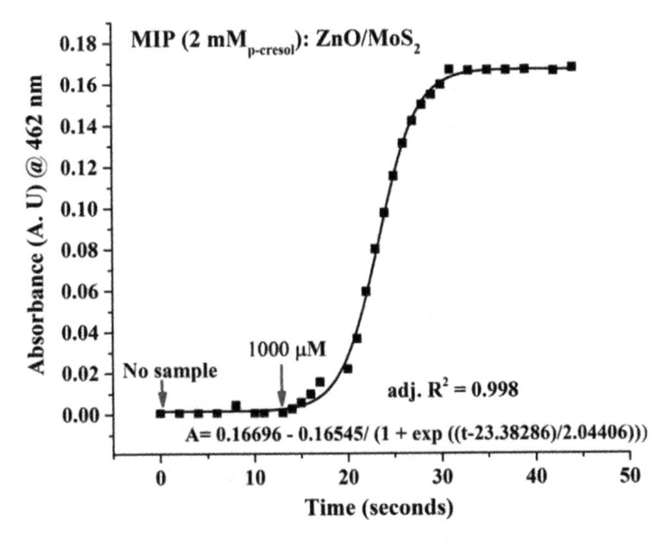

FIGURE 9.10 Response time of LMR-based p-cresol sensor. Source: *Figure reprinted from S.P. Usha, B.D. Gupta, Urinary p-cresol diagnosis using a nanocomposite of ZnO/MoS$_2$ and molecular imprinted polymer on optical fiber-based lossy mode resonance sensor, Biosens. Bioelectron. 101, 135–145 (2018) with permission from Elsevier.*

The time taken by a sensor after use to be ready for the next measurement is called the recovery time. A sensing device should rest immediately or after one measurement to resume its base equilibrium. The recovery of the sensor should be fast or it should have a small recovery time.

9.2.8.2 Lifetimes

The response of the sensor during continuous use of its probe starts decreasing. The time after which the response has declined by a given percentage is called the "lifetime" of the sensor. On the other hand, it is the time over which the assembled sensor is stored. The lifetime of the probe depends on a number of factors such as the stability of chemicals used for its fabrication because with time their activity deteriorates. The other factor is the deterioration of sensing and other layers of the probe. With time the peeling of layers from the substrate starts.

9.3 Summary

In this chapter, we have discussed the performance parameters such as sensitivity, selectivity, the limit of detection, and the response time of various types of metal chalcogenide-based biosensors. To explain these parameters, a number of examples have been taken from various studies reported in the literature.

References

[1] S. Kaushik, U.K. Tiwari, A. Deep, R.K. Sinha, Two-dimensional transition metal dichalcogenides assisted biofunctionalized optical fiber SPR biosensor for efficient and rapid detection of bovine serum albumin, Sci. Rep. 9 (2019) 6987.

[2] J. Lu, X. Zhang, N. Liu, X. Zhang, Z. Yu, T. Duan, Electrochemical detection of Cu^{2+} using graphene–SnS nanocomposite modified electrode, J. Electroanalyt. Chem. 769 (2016) 21–27.

[3] T. Lin, L. Zhong, Z. Song, L. Guo, H. Wu, Q. Guo, et al., Visual detection of blood glucose based on peroxidase-like activity of WS_2 nanosheets, Biosens. Bioelectron. 62 (2014) 302–307.

[4] C. Zhao, X. Wu, X. Zhang, P. Li, X. Qian, Facile synthesis of layered CuS/RGO/CuS nanocomposite on Cu foam for ultrasensitive nonenzymatic detection of glucose, J. Electroanalyt. Chem. 785 (2017) 172–179.

[5] J. Huang, Z. Dong, Y. Li, J. Li, W. Tang, H. Yang, et al., MoS_2 nanosheet functionalized with Cu nanoparticles and its application for glucose detection, Mater. Res. Bull. 48 (2013) 4544–4547.

[6] A. D'Amico, C.D. Nitale, A contribution on some basic definitions of sensors properties, IEEE Sens. J. 3 (2001) 183–190.

[7] K. Danzer, Selectivity and specificity in analytical chemistry. General consideration and attempt of a definition and quantification, Fresenius J. Anal. Chem. 369 (2001) 397–402.

[8] J. Vessman, Selectivity or specificity? Validation of analytical methods from the perspective of an analytical chemist in the pharmaceutical industry, J. Pharm. Biomed. Anal. 14 (1996) 867–869.

[9] D.G. Boef, A. Hulanicki, Recommendations for the usage of selective, selectivity and related terms in analytical chemistry, Pure Appl. Chem. 55 (1983) 553–556.

[10] S. Sharma, A.M. Shrivastav, B.D. Gupta, Lossy mode resonance based fiber optic creatinine sensor fabricated using molecular imprinting over nanocomposite of MoS_2/SnO_2, IEEE Sens. J. 20 (2020) 4251–4259.

[11] Q. Feng, K. Duan, X. Ye, D. Lu, Y. Du, C. Wang, A novel way for detection of eugenol via poly(diallyldimethylammonium chloride) functionalized graphene-MoS_2 nano-flower fabricated electrochemical sensor, Sens. Actuat. B 192 (2014) 1–8.

[12] G.L. Long, J.D. Winefordner, Limit of detection: a closer look at the IUPAC definition, Anal. Chem. 55 (1983) 712A–724A.

[13] A. Shrivastava, V.B. Gupta, Methods for the determination of limit of detection and limit of quantification of the analytical methods, Chron. Young Sci. 2 (2011) 21–25.

[14] S.P. Usha, B.D. Gupta, Urinary p-cresol diagnosis using nanocomposite of ZnO/MoS_2 and molecular imprinted polymer on optical fiber based lossy mode resonance sensor, Biosens. Bioelectron. 101 (2018) 135–145.

CHAPTER

10

Immunosensors and recommendations

10.1 Immunosensors

A special design of biosensors that incorporates a biological diagnosis mechanism with a transducer is named "Immunosensor." This device produces an assessable signal which is proportional to the variation of defined biomolecule concentration.

Cadmium sulfide as an important metal chalcogenide plays the role of a photoactive substance in the fabrication of the following immunosensors.

Determination of Human immunoglobulin antigen (H-IgG) concentration as a probable cancer biomarker is very vital.

Liu et al. designed an immunosensor with a signal-off strategy basis. In the device, TiO_2 is coupled with CdS quantum dots (QDs) and applied as the photoactive matrix and copper (II) ion (Cu^{2+}) as an inhibitor. This coupling promoted sensitivity. Indium tin oxide (ITO) electrode improved with TiO_2/CdS and was applied for primary antibody (Ab_1) immobilization and the subsequent sandwich-type antibody–antigen (Ab–Ag) affinity interactions. Specific affinity interactions between Ab_2 and Ag immobilized the flower-like copper oxide (CuO) on the improved electrode. CuO particles were labeled with secondary antibody (Ab_2) (Scheme 10.1).

In the photoelectrochemical (PEC) biosensor, produced Cu^{2+} (from the dissolution of CuO in HCl) reacted with cadmium sulfide and yielded copper sulfide Cu_xS ($X = 1, 2$). The Cu_xS has novel energy levels for electron–hole recombination and a decrease in the photocurrent outcomes. The addition of polythiophene (PT-Cl) on the surface of TiO_2 caused the PEC signal more stable. The immunosensor showed linear

DOI: https://doi.org/10.1016/B978-0-323-85381-1.00003-9

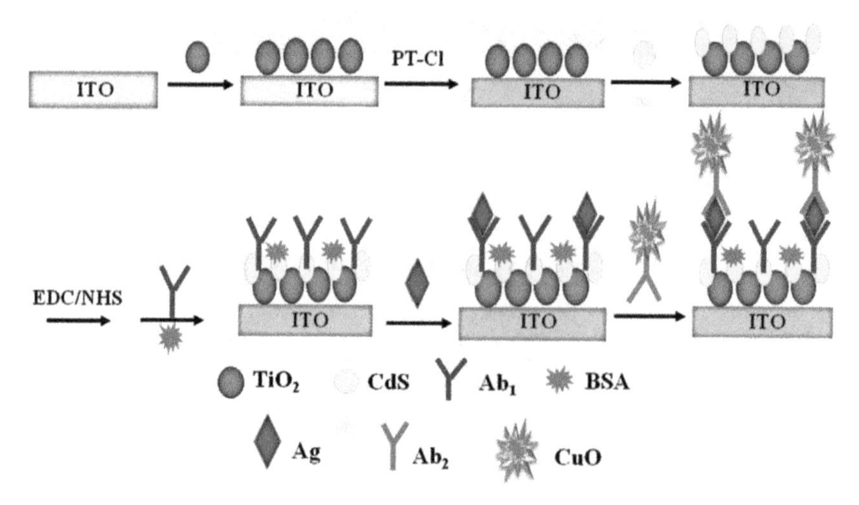

SCHEME 10.1 Fabrication process of the Liu et al. immunosensor.

range from 0.1 pg/mL to 100 ng/mL with a low detection limit of 0.03 pg/mL for (H-IgG) determination (405 nm irradiation at 0.1 V) [1].

Zeng et al. applied hollow cadmium sulfide(H-CdS) in an immunosensor for CarcinoEmbryonic Antigen (CEA) detection. H-CdS and CEA are photoactive matrices and tumor biomarkers respectively. Photocurrent intensity of hollow structure was higher than nanoparticle structure in CdS.

When target CEA was added, anti-C = EA capture antibody-conjugated immunomagnetic bead (IMB) reacted with horseradish peroxidase (HRP)-labeled anti-CEA detection antibody. The product of the reaction was the immunocomplex of the IMB antibody/CEA/HRP-labeled antibody (IMB-CEA-HRP). The carried HRP on the sandwich structure triggers enzymatic etching of H-CdS was magnetically separated. Therefore photocurrent intensity was reduced (Scheme 10.2).

Photocurrent responses toward target CEA were in the range of 0.02 to 50 ng/mL at a low detection limit of 6.12 pg/mL. Three important advantages of this method are high precision, great anti-interference ability, and acceptable accuracy. Based on this research, the strategy of hollow-nanostructure CdS activity and technology of enzyme-triggered etching was developed [2].

Zhu et al. designed a method for the fabrication of 3D CdS nanosheet (NS)-enwrapped carbon fiber framework (CFF). They used the method for sensitive split-type CuO-mediated PEC immunoassay. The 3D CdS NS-CFF was prepared by a solvothermal process. The sandwich immunocomplex was permitted in a 96-well plate with CuO NPs as the signaling labels. The liberated copper(II) ions(Cu^{2+} ions) interacted with the CdS NS

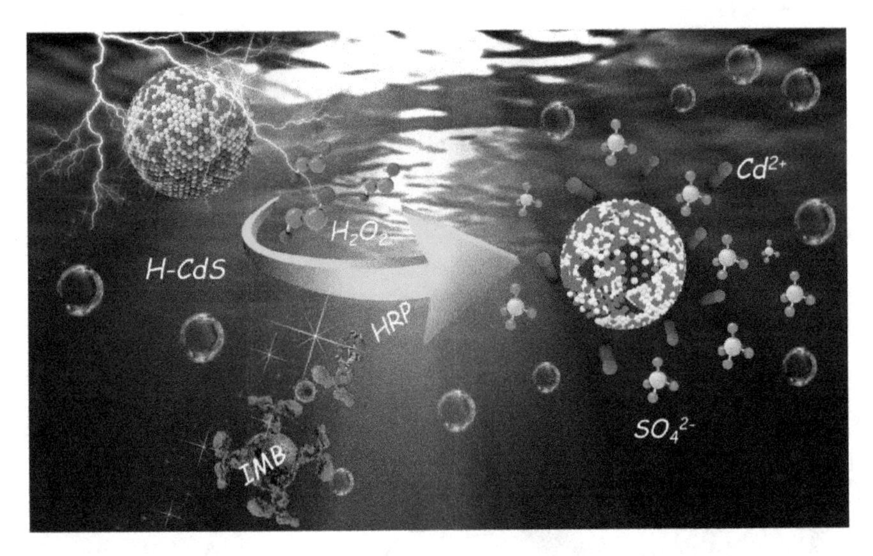

SCHEME 10.2 Schematic exhibition of carcinoembryonic antigen (CEA) detection on anti-CEA capture antibody-conjugated IMB (immunomagnetic bead).

and trapping sites produced, thus inhibiting its photocurrent generation. The 3D CdS NS-CFF photoelectrode was acting to guarantee its sufficient contact with the Cu^{2+} -containing solution and provide plenty of CdS surface for the Cu^{2+} ions (Scheme 10.3).

Liberation of the Cu^{2+} ions was dependent on the target. The copper ions are coupled with the 3D CdS NS-CFF photoelectrode properly. Therefore, a sensitive split-type PEC immunosensor for recognition of brain natriuretic peptide (BNP) has resulted.

The typical usability of the biosensor was proved when real sample analysis was compared with the commercial BNP enzyme-linked immunosorbent assay (ELISA) kit [3].

Recently, a sandwich-type PEC immunosensor was fabricated for PSA (prostate-specific antigen) detection. In this apparatus was applied ternary CdS@Au-g-C_3N_4 heterojunction as a photoelectrochemical matrix. Also, gold nanoparticles were used as photosensitizers and electron relays to improve the PEC performance (Scheme 10.4). On the other hand, GO-CuS could considerably quench the photocurrent from CdS@Au-g-C_3N_4 [4].

In 2020, a new ultrasensitive and selective Prion (PrP^C) immunosensor was devised using a potential-induced photocurrent-direction switching of the CdS-chitosan nanoparticles (CdS-CS NPs) and a direct Z-scheme CdS/hemin photocurrent-direction switching system [5].

The CdS-CS NPs served as photoactive materials that demonstrated a cathodic photocurrent by the potential-stimulated switching of the photocurrent direction strategy. Since immobilization of the captured

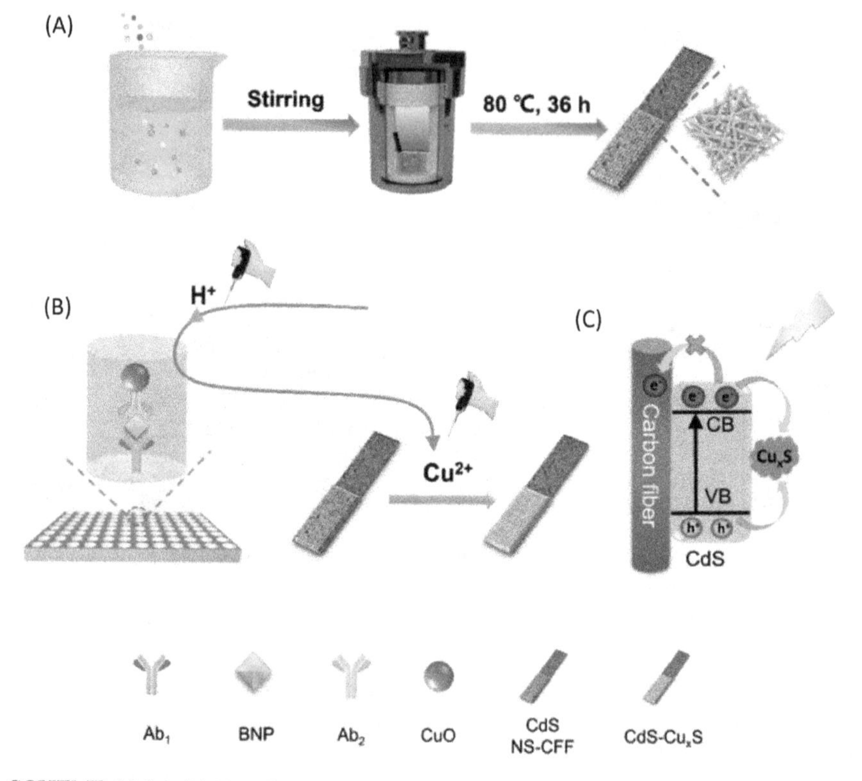

SCHEME 10.3 (A) Manufacture steps of 3D CdS NS-CFF electrode. (B) The electrode application for CuO-mediated PEC immunoassay. (C) The mechanism of photocurrent quenching.

antibody of PrP^C (Ab1) over the surface of CdS-CS NPs, nitrogen-doped porous carbon (NPC)-hemin polyhedra labeled with the secondary antibody of PrPC (Ab_2). The Ab_2 were subsequently introduced onto the immunosensing interface due to the antigen–antibody specific recognition and could act as a photocurrent-direction converter of CdS-CS NPs to yield a great anodic photocurrent which was against that of CdS-CS NPs, because of the well-matched energy levels between CdS and hemin to appear a direct Z-scheme CdS/hemin system (Scheme 10.5).

In the sensitive PEC immunosensor, PrP^C was detected with a wide linear response range from 4 to 1000 aM and a low detection limit of 0.53 Am and illuminated that the strategy may have promising applications in bioanalysis, disease diagnostics, and clinical biomedicine [5].

In recent years, an insulin biosensor designed by using C-TiO_2/CdS sensitized structure and PEC performance. For signal amplification labels of the immunosensor used from CuS-SiO_2 composites (Scheme 10.6). The competition effect of CuS and steric hindrance caused the synergistic effect in this device [6].

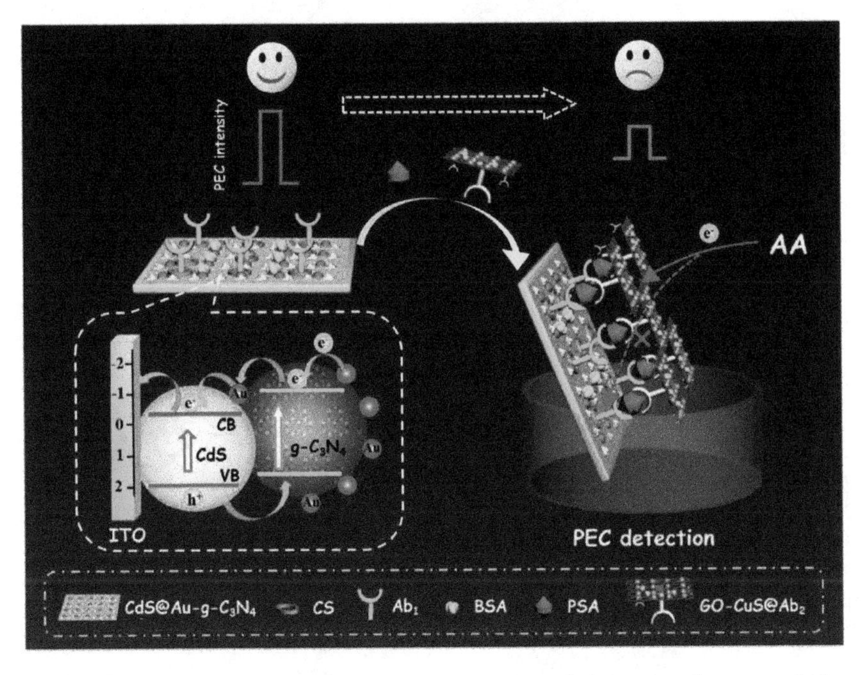

SCHEME 10.4 Schematic of the PEC immunosensor and charge transfer process before and after immobilization with GO-CuS@Ab_2 signal tag.

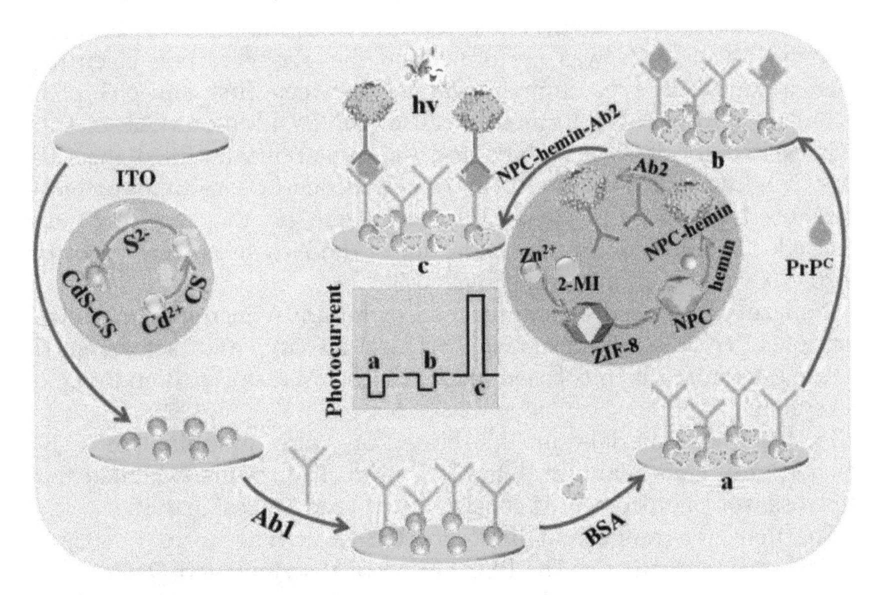

SCHEME 10.5 Schematic illumination for a PEC immunosensor accompanying a direct Z-scheme CdS/hemin system.

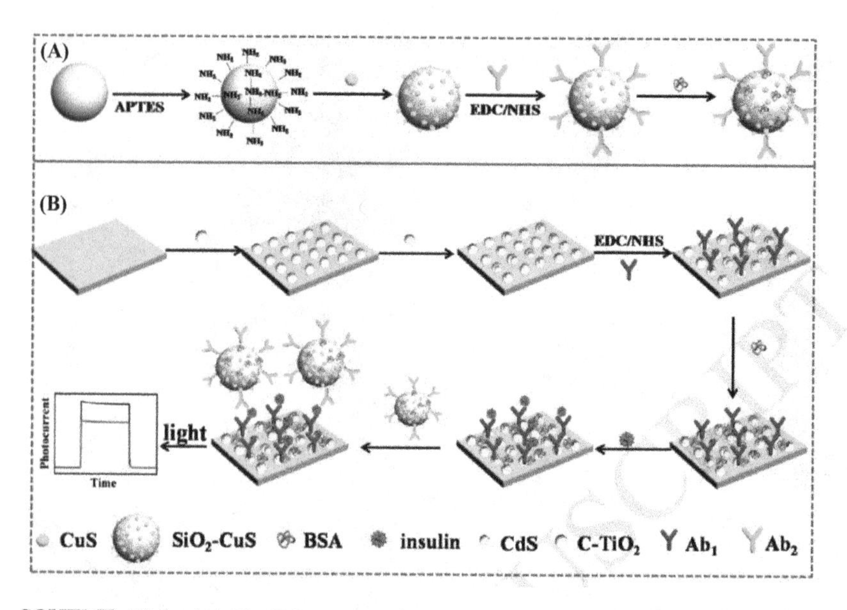

SCHEME 10.6 (A) The Fabrication of the bioprobe. (B) The construction steps of the PEC immunosensor.

Heidari et al. designed an Electrochemiluminescence (ECL) immunosensor for selective detection of a cancer biomarker:p53 protein. Sensing of the protein was based on the enhancement effects of AuNPs over ECL emission of CdS nanocrystals (CdS NCs). In the first step, immobilization of CdS NCs was performed on the glassy carbon electrode. Then a sandwich type immunocomplex between first anti-p53/p53/ secondary anti-p53 was formed to cause AuNPs addition. ECL of CdS NCs excited the SPR of AuNPs and the excitation amplified the CdS NCs ECL intensity (Scheme 10.7). To immunosensor sensitivity increase were used graphene oxide and AuNPs were on the surface of the electrode. The detection range and limit of the biosensor were reported as 20–1000 and 4 fg/mL, respectively [7].

Wang et al. devised an immunosensor for DNA methylation, detection of DNA methyltransferase (MTase) activity, and screening of MTase inhibitor, which is based on the M. SssI MTase-HpaII endonuclease system.

Photoactive materials in the biosensor were graphite-like C_3N_4 (g-C_3N_4) and CdS quantum dots (CdS QDs). The results exhibited that the biosensor photocurrent is consistent with the M. SssI activity.

The detection range and limit of M. SssI activity were 1–80 and 0.316 U/mL respectively. The PEC biosensor discriminated DNA methylation, detected DNA MTase activity, and recognized its inhibitor based on DNA methylation triggering the inhibition of HpaII activity

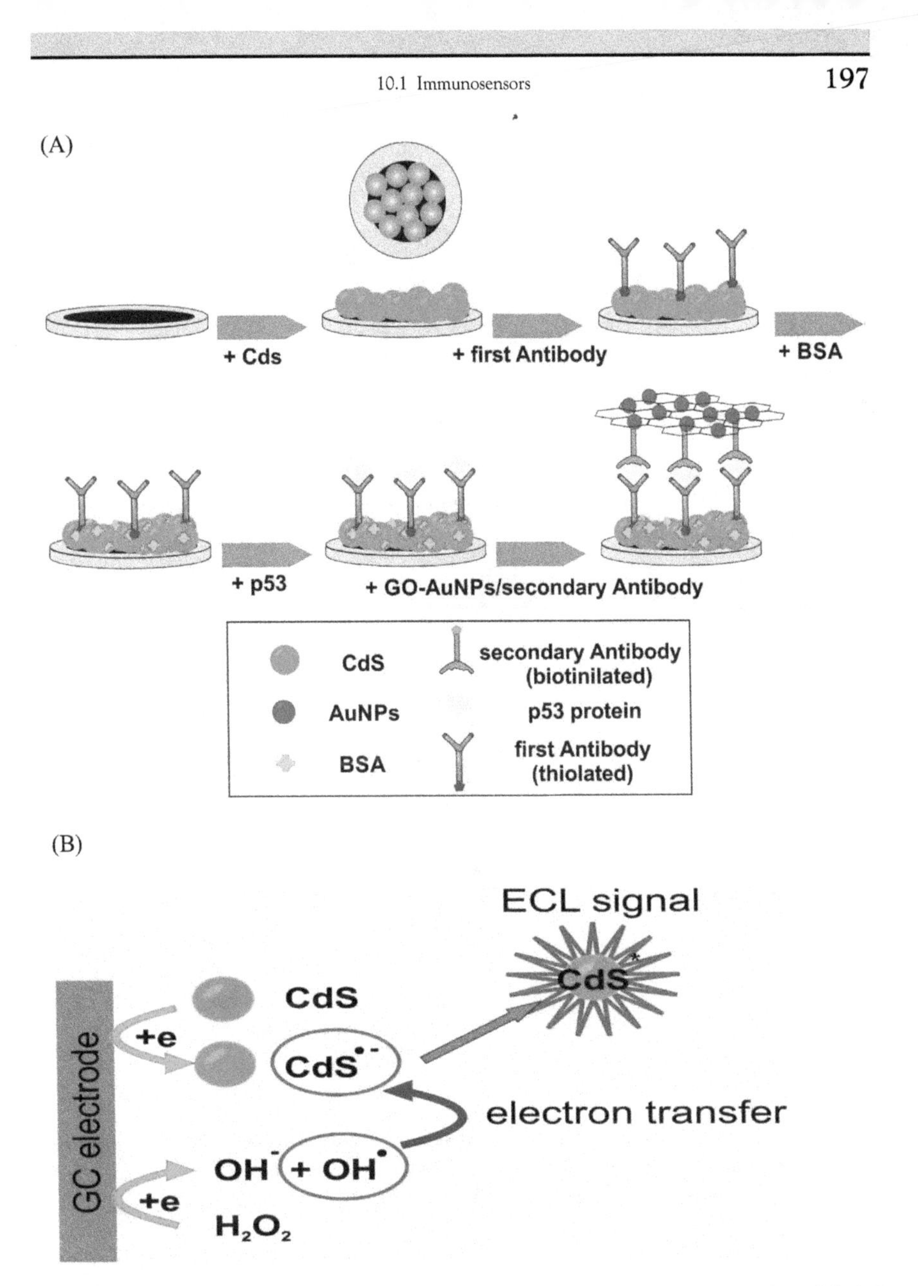

SCHEME 10.7 (A) ECL immunosensor applying CdS NCs and tGO AuNPs (B) ECL signal production of CdS in the presence of H_2O_2 Improper DNA methylation is a pre-indicator of various diseases.

and the improved PEC property of g-C_3N_4 by CdS QDs (Scheme 10.8). The immunosensor can be used for DNA MTase inhibitors detection because azamethipos (a kind of pesticide) influences M. SssI MTase activity [8].

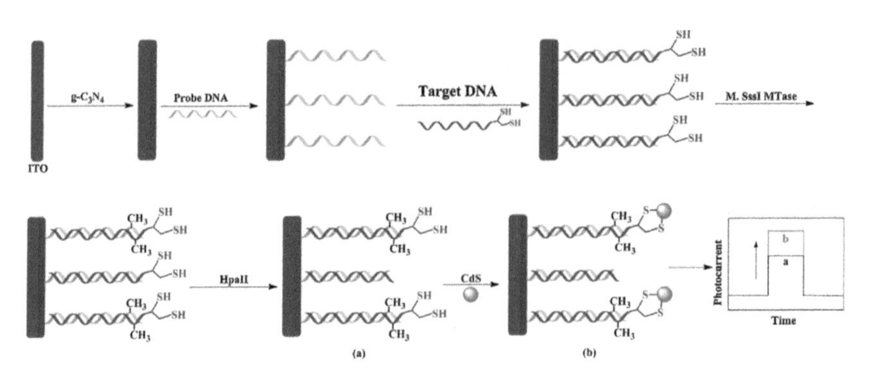

SCHEME 10.8 Illustration of the PEC biosensor for M. SssI MTase activity assay.

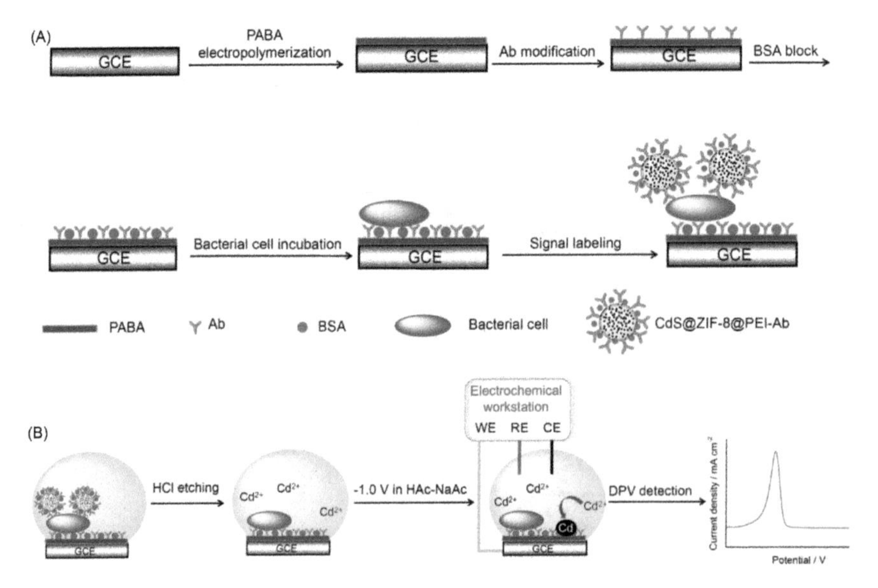

SCHEME 10.9 (A) Illustration of the construction steps of the sandwich-type electrochemical immunobiosensor for the quantification of *Escherichia coli* O157:H7 using CdS@ZIF-8 as signal tags. (B) Illustration of the steps for DPV assay.

For the detection of *Escherichia coli* O157:H7 (*E. coli* O157:H7), two immunosensors were designed. The first was by Zhong et al., that in the device, CdS@ZIF-8 muti-core-shell particles were prepared by in situ growth of ZIF-8 in attendance of CdS QDs and applied as signal-amplifying tags [9]. Fabrication and detection steps are illustrated in Scheme 10.9.

The second by Dong et al., that in the biosensor Au NPs and CdS QDs were assembled on the aligned ZnO nanowire to fabricate a self-powered PEC platform. This initial platform has favorable energy band

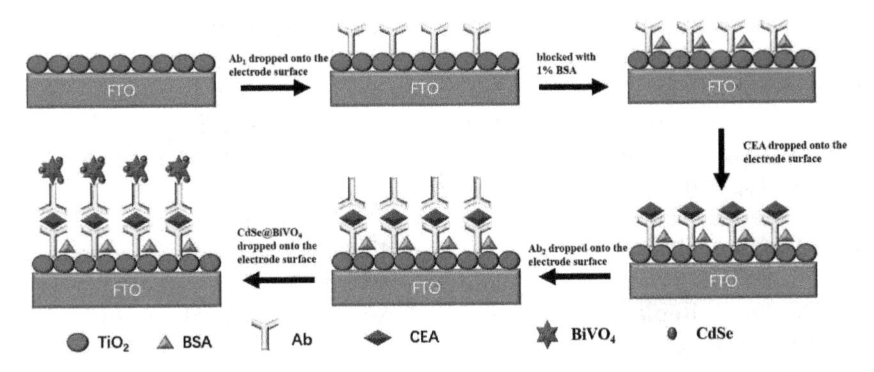

SCHEME 10.10 The fabrication steps of the CEA immunosensor. *CEA*, CarcinoEmbryonic Antigen.

configuration and further developed into a sensitive aptasensor. Three factors for development are merging the nice electron transfer route of ZnO, the extended visible-light absorption, and the localized surface plasmon of Au NPs [10].

Im et al. fabricated an immune sensor for horseradish peroxidase and luminol detection and applied it to the medical detection of carcinoembryonic antigen. In the photosensor applied CdS NW with the perylene-C passivation layer [11].

Xie et al. constructed an ultrasensitive photoelectrochemical (PEC) immunosensor for the detection of Carcinoembryonic Antigen (CEA). Photoactive materials of the biosensor were CdSe@$BiVO_4$ co-sensitized TiO_2 nanorods. To immobilization of capture antibodies, TiO_2 nanorods were assembled on the FTO-modified electrode. Three determinative aspects of sensitivity are as follows:(1) the co-sensitization of $BiVO_4$ and CdSe broadens the absorption range of TiO_2 from the ultraviolet region to the visible light region, which can use light energy, (2) the effective matching of energy levels through TiO_2, CdSe and $BiVO_4$ accelerates the dissociation and transition of photogenerated electron–hole pairs and promotes the PEC performance, (3) the added Au advances the interfacial electron transfer from TiO_2 to FTO electrode (Scheme 10.10). At optimized conditions, linear detection range from 0.01–50 ng/mL and a detection limit of 0.5 pg/mL were observed [12].

10.2 Recommendations for future research

Several remarkable points are recommended for the previous chapters.

Chapter 3: The tables shown in the chapter can be a useful guide for designing the synthesis and preparation of metal chalcogenides used in the construction of biosensors. Empty cells in these tables predict several metal chalcogenides. For example, in Table 3.12. Shown for the preparation of quaternary metal chalcogenides, three types of metal chalcogenides and variable molar ratios can be used. Empty cells can pave the way for designing the synthesis and molar ratio probabilities of three types of metal chalcogenides.

Chapter 4: The diversity of newly explored gaseous products produced in biological processes requires further research to build advanced biosensors using metal chalcogenides.

Chapter 5: Replacing existing metal oxide in chemical sensors with available synthetic metal chalcogenides can significantly increase the efficiency of chemical biosensors [13].

Chapter 7: In Table 7.1, four components (metal chalcogenide-electrode-complement-method) are proposed to prepare a protein biosensor. Vacant cells in the table and similar tables that could be designed demonstrate the way for a more accurate biosensor in the future. Kind of the bioanalyte that fill vacant cells depends on previous experiments and obtained result.

Chapter 8: Use new tools, techniques, and materials in manufacturing and production, such as 3D printing technology, nanomaterials, and up-to-date microchips in the construction of biosensors [14,15,16]. The appropriate concentration range of bio analyte in the biosensor is approximated by software such as COMSOL [17]. Applying advanced simulation software help to accurately the performance of biosensor considerably.

References

[1] Y. Liu, R. Li, P. Gao, Y. Zhang, H. Ma, J. Yang, et al., A signal-off sandwich photoelectrochemical immunosensor using TiO2 coupled with CdS as the photoactive matrix and copper (II) ion as inhibitor, Biosens. Bioelectron. 65 (2015) 97–102. Available from: https://doi.org/10.1016/j.bios.2014.10.020.

[2] R. Zeng, D. Tang, Magnetic bead-based photoelectrochemical immunoassay for sensitive detection of carcinoembryonic antigen using hollow cadmium sulfide, Talanta (2020) 219. Available from: https://doi.org/10.1016/j.talanta.2020.121215.

[3] Y.C. Zhu, Z. Li, X.N. Liu, G.C. Fan, D.M. Han, P.K. Zhang, et al., Three-dimensional CdS nanosheet-enwrapped carbon fiber framework: towards split-type CuO-mediated photoelectrochemical immunoassay, Biosens. Bioelectron. 148 (2020). Available from: https://doi.org/10.1016/j.bios.2019.111836.

[4] J.T. Cao, Y.X. Dong, Y. Ma, B. Wang, S.H. Ma, Y.M. Liu, A ternary CdS@Au-g-C3N4 heterojunction-based photoelectrochemical immunosensor for prostate specific antigen detection using graphene oxide-CuS as tags for signal amplification, Anal. Chim. Acta 1106 (2020) 183–190. Available from: https://doi.org/10.1016/j.aca.2020.01.067.

[5] R. Yang, J. Liu, Sensitive and selective photoelectrochemical immunosensing platform based on potential-induced photocurrent-direction switching strategy and a direct Z-scheme CdS//hemin photocurrent-direction switching system, J. Electroanal. Chem. 873. (2020). Available from: https://doi.org/10.1016/j.jelechem.2020.114346

[6] X. Wang, P. Gao, T. Yan, R. Li, R. Xu, Y. Zhang, et al., Ultrasensitive photoelectrochemical immunosensor for insulin detection based on dual inhibition effect of CuS-SiO2 composite on CdS sensitized C-TiO2, Sens. Actuators B Chem. 258 (2018) 1–9. Available from: https://doi.org/10.1016/j.snb.2017.11.073.

[7] R. Heidari, J. Rashidiani, M. Abkar, R.A. Taheri, M.M. Moghaddam, S.A. Mirhosseini, et al., CdS nanocrystals/graphene oxide-AuNPs based electrochemiluminescence immunosensor in sensitive quantification of a cancer biomarker: p53, Biosens. Bioelectron. 126 (2019) 7–14. Available from: https://doi.org/10.1016/j.bios.2018.10.031.

[8] H. Wang, P. Liu, W. Jiang, X. Li, H. Yin, S. Ai, Photoelectrochemical immunosensing platform for M. SssI methyltransferase activity analysis and inhibitor screening based on g-C3N4 and CdS quantum dots, Sens. Actuators B Chem. 244 (2017) 458–465. Available from: https://doi.org/10.1016/j.snb.2017.01.016.

[9] M. Zhong, L. Yang, H. Yang, C. Cheng, W. Deng, Y. Tan, et al., An electrochemical immunobiosensor for ultrasensitive detection of Escherichia coli O157:H7 using CdS quantum dots-encapsulated metal-organic frameworks as signal-amplifying tags, Biosens. Bioelectron. 126 (2019) 493–500. Available from: https://doi.org/10.1016/j.bios.2018.11.001.

[10] X. Dong, Z. Shi, C. Xu, C. Yang, F. Chen, M. Lei, et al., CdS quantum dots/Au nanoparticles/ZnO nanowire array for self-powered photoelectrochemical detection of Escherichia coli O157:H7, Biosens. Bioelectron. 149 (2020). Available from: https://doi.org/10.1016/j.bios.2019.111843.

[11] J.H. Im, H.R. Kim, B.G. An, Y.W. Chang, M.J. Kang, T.G. Lee, et al., In situ-synthesized cadmium sulfide nanowire photosensor with a parylene passivation layer for chemiluminescent immunoassays, Biosens. Bioelectron. 92 (2017) 221–228. Available from: https://doi.org/10.1016/j.bios.2017.02.021.

[12] Y. Xie, M. Zhang, Q. Bin, S. Xie, L. Guo, F. Cheng, et al., Photoelectrochemical immunosensor based on CdSe@BiVO4 Co-sensitized TiO_2 for carcinoembryonic antigen, Biosens. Bioelectron. 150 (2020). Available from: https://doi.org/10.1016/j.bios.2019.111949.

[13] M.A. Abdelkareem, E.T. Sayed, A. Iqbal, C. Rodriguez, A.-G. Olabi, Progress in the use of metal chalcogenides for batteries, in: A.-G. Olabi (Ed.), Encyclopedia of Smart Materials, Elsevier, 2022, pp. 166–175. ISBN9780128157336. Available from: https://doi.org/10.1016/B978-0-12-815732-9.00102-9.

[14] M. Habibi, P. Bagheri, N. Ghazyani, H. Zare-Behtash, E. Heydari, 3D printed optofluidic biosensor: $NaYF^4$: Yb^{3+}, Er^{3+} up conversion nano-emitters for temperature sensing, Sens. Actuators A Phys. 326 (2021) 112734. Available from: https://doi.org/10.1016/j.sna.2021.112734. ISSN0924-4247.

[15] J. Muñoz, M. Pumera, 3D-printed biosensors for electrochemical and optical applications, TrAC Trends Analyt. Chem. 128 (115933) (2020) ISSN 0165-9936. Available from: https://doi.org/10.1016/j.trac.2020.115933.

[16] Y. Zhang, Xianzhi Hu, Qingjiang Wang, Y. Zhang, Recent advances in microchip-based methods for the detection of pathogenic bacteria, Chin. Chem. Lett. (2021) ISSN 1001-8417. Available from: https://doi.org/10.1016/j.cclet.2021.11.033.

[17] N.P. Nambiar, C. Paul, G. Girish, A. Velayudhan, S.D. Baby Sreeja, P.R. Sreenidhi, Simulation study of micro fluidic device for biosensor application, Materials Today: Proceedings 46 (2021) 3158–3163. Available from: https://doi.org/10.1016/j.matpr.2021.03.247.

Further reading

M. Bouroushian, Electrochemistry of Metal Chalcogenides, Springer Berlin Heidelberg, Germany, 2010. ISBN: 978-3-642-03967-6.

R. K. Gupta, T. Rasheed, T. A. Nguyen, M. Bilal Metal-Organic Frameworks-Based Hybrid Materials for Environmental Sensing and Monitoring, CRC Press, N.p., 2022. ISBN 9781032024530.

A. Pandikumar, P. Rameshkumar, Graphene-Based Electrochemical Sensors for Biomolecules, Elsevier Science, Netherlands, 2018, Paperback ISBN: 9780128153949, eBook ISBN: 9780128156391.

Index

Note: Page numbers followed by "*f*" and "*t*" refer to figures and tables, respectively.

CPSIA information can be obtained
at www.ICGtesting.com
Printed in the USA
BVHW050526280123
657287BV00011B/306

9 780323 853811